13.75

MAN IN EXTREME ENVIRONMENTS

Publication Number 1033

AMERICAN LECTURE SERIES®

A Monograph in

The BANNERSTONE DIVISION *of*
AMERICAN LECTURES IN ENVIRONMENTAL STUDIES

Edited by

CHARLES G. WILBER

Chairman and Professor of Zoology
Colorado State University
Fort Collins, Colorado

MAN IN EXTREME ENVIRONMENTS

By

A. W. SLOAN, M.D., Ph.D., F.R.C.P. (Glasgow), M.R.C.P. (London), F.R.S. (South Africa)

*Professor and Head
Department of Physiology
University of Cape Town
Cape Town, South Africa*

CHARLES C THOMAS • PUBLISHER
Springfield • Illinois • U.S.A.

Published and Distributed Throughout the World by
CHARLES C THOMAS ● PUBLISHER
Bannerstone House
301-327 East Lawrence Avenue, Springfield, Illinois, U.S.A.

© *1979, by* CHARLES C THOMAS ● PUBLISHER
ISBN 0-398-03941-0
Library of Congress Catalog Card Number: 79-13940

With THOMAS BOOKS *careful attention is given to all details of
manufacturing and design. It is the Publisher's desire to present books that
are satisfactory as to their physical qualities and artistic possibilities and
appropriate for their particular use.* THOMAS BOOKS *will be true to those
laws of quality that assure a good name and good will.*

Printed in the United States of America
V-R-1

Library of Congress Cataloging in Publication Data

Sloan, Archibald Walker.
 Man in extreme environments.

 (American lecture series ; no. 1033)
 Bibliography: p. 111
 Includes index.
 1. Adaptation (Physiology) 2. Man--Influence of
environment. I. Title.
QP82.S54   612'.0144   79-13940
ISBN 0-398-03941-0

FOREWORD

THE profound influence that environmental factors have on man and his endeavors has been recognized since ancient times. Hippocrates in *Air, Waters, Places* exhorts the medical scientists of his times as follows:

> Whoever wishes to pursue properly the science of medicine must proceed thus. First he ought to consider what effects each season of the year can produce; for the seasons are not at all alike, but differ widely both in themselves and at their changes. The next point is the hot winds and the cold, especially those that are peculiar to each particular region. He must also consider the properties of the waters; for as these differ in taste and in weight, so the property of each is far different from that of any other. Therefore, on arrival at a town with which he is unfamiliar, a physician should examine its position with respect to the winds and to the risings of the sun. For a northern, a southern, an eastern, and a western aspect has each its own individual property. He must consider with the greatest care both these things and how the natives are off for water, whether they use marshy, soft waters, or such as are hard and come from rocky heights, or brackish and harsh. The soil too, whether bare and dry or wooded and watered, hollow and hot or high and cold. The mode of life also of the inhabitants that is pleasing to them, whether they are heavy drinkers, taking lunch [that is, taking more than one full meal every day], and inactive, or athletic, industrious, eating much and drinking little.

The competence of such advice is today recognized by most biomedical scientists.

In any discussion of human adaptation to environment, it is essential to appreciate that physiological adaptation includes a wide spectrum of phenomena of physiology, and indeed bears upon general principles of biology (Dill, Edwards, and Wilber, 1964).

Simple all-encompassing explanations of man's adaptations to his surroundings in most instances are misleading and at best inadequate. Cabanac (1975) emphasizes this fact when he concludes that in man "temperature regulation is brought about by multiple independent feedback loops." The complete system involves an adjustable set point and proportional control.

Man cannot get out of contact with his surroundings nor can he ignore environmental forces (Wilber, 1958). They act on him individually and as an animal species (population effects).

There are voluminous records of research available on the impact of environment on man, but many crucial questions are left unanswered. Sloan in this monograph has put together a "state of the art" summary which covers some fundamental biological mechanisms of adaptation in man followed by specific analyses of human adaptations to heat, cold, deep diving conditions, montane environments, and then the man-made environmental conditions forced upon certain human beings involved in aviation and in space flight.

The problem of purely morphological adjustments in man as an effective adaptation to cold still seems to be debated. A number of reports emphasize the thermodynamic ineffectiveness of body shape changes in man in increasing his survival chances in the cold (Wilber, 1957; Scholander, 1955; Hicks, 1964; for example). There seems to be a trend, however, that reveals technological creativity in some human varieties has permitted those groups to thrive in the cold, whereas in other human varieties, specific, centrally mediated physiological mechanisms have been identified as unique, adaptive human responses to cold.

Sloan shows clearly the responses that *Homo sapiens* has made and will continue to make (both biological and technological) resulting in man as ubiquitous on planet Earth and perhaps eventually in the sun's region of the universe.

A relatively new challenge faces our species — the adaptation of man to his own creativity. Human creative output includes a bewildering assortment of synthetic chemicals, noise, machines, person to person interactions, voluminous and suffocating devices of "communication" (infinite computer printouts, television, radio, new sources of radioactivity, recorders and playback devices for voice records, for example). This complex deluge has yet to be mastered by man. Inevitably, man must and will adapt to these environmental insults to his own created by himself.

Research on these new man-made environmental stressors is an exciting and imperative challenge for young scientists. The existence of these stressors gives the lie to those few prognosticators who have implied that scientific endeavor is near its end and that all the really important scientific discoveries have been made.

We can read Sloan's book and quickly realize that the present state of knowledge about man's adaptation to environmental extremes proves the past research successes of biomedical scientists and points unequivocally to future challenges of critical importance — but challenges that are bathed in the clear light of hope.

Fort Collins, Colorado   Charles G. Wilber

REFERENCES

Cabanac, M.: Temperature regulation. *Ann Rev Physiol, 37*:415-439, 1975.

Dill, D. B., Edwards, E. A. and Wilber, C. G. (Eds.): *Adaptation to the Environment*. Washington, D. C., American Physiological Society, 1964, 1056 pp.

Hicks, C. S.: Terrestrial animals in cold; exploratory studies in primitive man. Dill, D. B., Edwards, E. A., and Wilber, C. G. (Eds.): In *Adaptation to the Environment*. Washington, D. C., American Physiological Society, 1964, pp. 405-412.

Scholander, P. F.: Evolution of climatic adaptation in homeotherms. *Evolution, 9*:15-26, 1955.

Wilber, C. G.: Physiological regulations and the origin of human types. *Hum Biol, 29*:329-336, 1957.

Wilber, C. G.: Man and his environment. *Armed Forces Chem J, 12*:12-13, 1958.

PREFACE

THIS monograph describes the stresses imposed on the human body by extreme environments, the normal response of the body to these stresses, the limits of adaptation, and the results of failure of adaptation. The environments studied include the tropics, the polar regions, deep water, and high altitudes. Since the principal stresses involved are due to high or low environmental temperature and high or low pressures of respiratory gases, the mechanisms of temperature regulation and of respiration are briefly described in the introductory chapters.

Because of the immediate danger to human life, when physiological adaptation fails in an extreme environment, signs of impending failure are described and the rationale of prevention and treatment of the commoner diseases due to environmental stress is explained. This knowledge is important even to people living in temperate climates, who may be caught unprepared by extreme weather conditions, to which people normally living in these conditions have some degree of acclimatization.

Environmental physiology is a vast subject and there is an extensive literature, including several textbooks, on each of the special environments described here. This book presents and explains current views on human adaptation to extreme environments; the experimental evidence for these views and relevant tables and graphs must be sought in the more specialized literature, to which reference is made. The list of references is necessarily selective and preference is given to review articles and books rather than to original reports, though the more important of these are included.

We are living today in a period of transition between traditional units of measurement, e.g. feet, pounds, and the metric units adopted in the Système International e.g. meters, kilograms. Traditional metric units are employed in this book with the equivalent other units in parenthesis where appropriate. Temperature is expressed in degrees Celsius (centigrade). Lists of equivalents and a conversion chart from the Fahrenheit to the Celsius scale of temperature are given in the Appendix.

I wish to thank Dr. Charles G. Wilber, editor of the AMERICAN LECTURES IN ENVIRONMENTAL STUDIES, who encouraged me to write this book, my colleagues at the University of Cape Town for constructive criticism of the text, Miss Jean Walker for the original illustrations, Mrs. Maureen Oosthuizen for typing the manuscript, and Mrs. Jean Sloan for help with the proofreading and for preparing the index. The authors and publishers, who have kindly permitted me to use their material for some of the figures, are acknowledged in the legends.

A. W. SLOAN

CONTENTS

MAN IN EXTREME ENVIRONMENTS

CHAPTER 1

ADAPTATION AND
TEMPERATURE REGULATION

ADAPTATION TO ENVIRONMENT

THE life of an animal or plant depends on continuous adaptation to changes in its environment, and the geographical distribution of animals and plants, under natural conditions, is restricted by their capacity for acclimatization. "Acclimatization" is a physiological change, occurring within the lifetime of an organism, which reduces the strain caused by stressful changes in the natural climate (Bligh and Johnson, 1973). By natural selection the organisms that can best adapt to environmental change are those that survive and propagate.

Man is the most adaptable of animals. He can live in the tropics, in the arctic, in deep caves, and on high mountains. With the aid of special equipment he can survive for prolonged periods even in the depths of the sea and in outer space. An individual who normally lives at low altitude in a temperate climate can live and work, with appropriate clothing and equipment, in all these different environments, but some racial adaptation is found, as a result of genetic selection, in populations that have lived for generations in an extreme environment.

In extreme environments *behavioral adaptation,* including wearing appropriate clothing, is at least as important as physiological adaptation and, in environments unnatural to a terrestrial animal (under water and in outer space), man can survive only with the aid of devices to maintain an immediate environment compatible with human life.

Although the mechanisms involved in human adaptation to environment are remarkably efficient, they can function only within a limited range of environmental changes. Beyond these limits, at extremes of cold, heat, pressure, or lack of oxygen, the compensatory mechanisms fail and illness or death results.

3

HOMEOSTASIS

The cells of a higher animal, such as man, can function normally only if the internal environment (the tissue fluids which surround and permeate the cells) is kept constant within fairly narrow limits. All the vital mechanisms, varied as they are, have only one object, that of keeping constant the conditions of life in the internal environment (Bernard, 1878). In homeothermic animals (including man) the temperature as well as the composition of the internal environment is kept constant in a wide range of temperatures of the external environment.

The maintenance of a constant internal environment, appropriately called "homeostasis" by Cannon (1939), depends on the integrated action of a number of mechanisms of adaptation, in which the nervous and endocrine systems play an important part. In general an immediate transient adaptation to a sudden change is regulated by the nervous system, whereas more sustained adaptation to prolonged change is a function of the endocrine system.

TEMPERATURE REGULATION

Body Temperature

Heat is produced as a by-product of metabolism in all living tissues and is lost from the skin and from the lungs. The body temperature depends on the balance of heat production and heat loss. The circulating blood plays a very important role in distributing heat fairly uniformly throughout the deep organs and in conveying it to the surface. It is convenient to distinguish the temperature of the deeper parts of the body (*core temperature*) from the temperature of the skin and subcutaneous tissue (*shell temperature*).

The core temperature in man is normally between 36°C and 38°C and varies about 1°C* in the course of 24 hours, being highest at about 6 PM and lowest at about 4 AM. The range of

---

*Conversion chart for Celsius (centigrade) and Fahrenheit scales of temperature in Appendix.

core temperature compatible with life is about 25 to 43°C (Herrington, 1949), but individuals have survived core temperatures higher or lower than this for varying periods of time with complete recovery (Burton and Edholm, 1955; Lee, 1964). Shell temperature is lower (mean about 33°C over the trunk) and varies considerably with different degrees of physical activity and in different environmental conditions. It falls progressively from the trunk to the extremities of the limbs; skin temperature at the finger may be 10°C lower than at the shoulder (Bell et al., 1976).

Measurement of Body Temperature

Body temperature is usually measured with a mercury-in-glass clinical thermometer held below the tongue with the lips closed (oral temperature) or inserted into the rectum (rectal temperature). Either of these measurements gives a rough indication of core temperature, though oral temperature is affected for some time by hot or cold food or drink and rectal temperature responds slowly to changes in core temperature and is influenced by bacterial activity in the rectum and by the temperature of the blood returning from the lower limbs. Immediate measure of the temperature of freshly passed urine (Fox et al., 1973) is a better indication of core temperature.

More sophisticated heat-sensitive devices are the thermocouple and the thermistor, either of which can be used to measure core or shell temperature. Core temperature is measured with the heat-sensitive element introduced into the oesophagus or near the tympanic membrane of the ear with the external auditory meatus blocked. The temperature of the tympanic membrane, which is adjacent to the internal carotid artery, is the best indication of the temperature of the blood going to the brain (Benzinger, 1969).

Regulation of Body Temperature

The balance between the rate of heat gain and the rate of heat loss is controlled by the temperature-regulating centers in the hypothalamus (Fig. 1). This part of the brain responds to changes in the temperature of the blood reaching it (core temperature) and to nerve impulses from temperature receptors in the skin and elsewhere by sending appropriate nerve impulses to the organs concerned with temperature regulation. The relative importance of the two modes of stimulation of the hypothalamus is still disputed (Cabanac, 1975). In general the anterior and medial groups of nuclei in the hypothalamus respond to warming of the blood or of the skin by increasing the rate of heat loss, whereas the posterior and lateral groups of nuclei respond to cooling by reducing the rate of heat loss and increasing the rate of heat production.

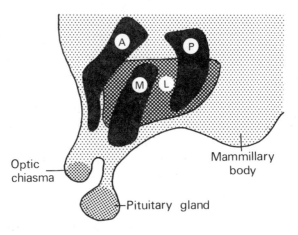

Figure 1. Groups of nuclei in the hypothalamus. Activity of A (anterior) and M (medial) increase heat loss; activity of L (lateral) and P (posterior) reduce heat loss and increase heat production.

Heat Loss

At normal and high environmental temperatures temperature regulation depends almost entirely on control of heat loss. Some loss of heat takes place from the lungs but most of it is from the skin (Table I).

Although a considerable amount of heat is lost through the lungs in a cool or temperate climate, this loss of heat is dependent on environmental factors and plays no part in the regulation of body temperature. The loss of heat depends on the temperature and humidity of the inspired air. Expired air is at body temperature and saturated with water vapor; the cooler and drier the inspired air the more heat is lost in raising it to body temperature and in the evaporation of water from the lungs.

TABLE I

ENERGY BALANCE FOR 24 HOURS
(arbitrary figures for sedentary man in temperate climate)

Intake	kcal	*Output*	kcal
In diet	2500	(1) As external work	200
		(2) As heat	
		Warming food and drink	50
		From lungs	
		(a) vaporisation of water	230
		(b) warming inspired air	100
		From skin	
		(a) radiation	900
		(b) convection	700
		(c) conduction	20
		(d) evaporation	300
	2500		2500

In a cool or temperate environment heat is lost from the skin by radiation, convection, conduction, and evaporation. The rate of heat loss by the first three of these physical principles depends on the temperature gradient between the skin and the external environment; heat is gained if the environment is warmer than the skin. A rise in shell temperature or a fall in environmental temperature increases the temperature gradient and the rate of heat loss. Shell temperature depends mainly on the rate of blood flow through the skin; the faster the flow of warm blood from the deeper parts of the body through the skin the higher is the shell temperature.

RADIATION. In a cool environment most heat loss from the

skin is by radiation. The human body is a very efficient radiator for the infrared (heat) waves of the solar spectrum. It readily absorbs radiant heat and as readily emits it, depending on the balance of radiant heat between the body and the environment.

CONVECTION. Loss of heat by convection depends on warming the thin layer of air next to the skin to the temperature of the skin. The warm air rises and its place is taken by cooler air. This mode of heat loss depends on skin temperature, air temperature, and air movement. The natural convection currents, which would cool the naked body in still air are accentuated by movement of air and reduced by clothing.

CONDUCTION. Loss of heat by conduction is small in air, because air is a poor conductor of heat, but much greater in water, when it may exceed loss by convection unless there is active movement through the water. The rate of heat loss by conduction depends on the temperature gradient between the skin and whatever is in contact with it, on the area of contact, and on the thermal conductivity of skin and of the substance in contact with it.

EVAPORATION. Even in a cool environment the body loses about 500 ml of water through the skin in 24 hours as insensible perspiration. Evaporation of this water accounts for loss of about 300 kcal of heat energy. In a humid environment evaporation from the skin is reduced and in an environment saturated with water vapor it does not occur. Loss of heat by evaporation is greatly increased by sweating, which may provide for a short period more than 3 liters/hour of water to evaporate from the surface of the skin (Ingram, 1977). To cool the skin the sweat must evaporate. Air movement aids evaporation, but a humid environment reduces it. In an atmosphere already saturated with water vapor sweat runs off the body without contributing to heat loss. When the environmental temperature is higher than the body temperature evaporation is the only mode of heat loss.

Regulation of Heat Loss

Physiological regulation of the rate of heat loss depends on control of cutaneous blood flow and on the secretion of sweat.

CUTANEOUS BLOOD FLOW. The rate of blood flow through the skin depends on the caliber of the cutaneous arterioles and arteriovenous anastomoses. Cooling the skin or the blood flowing to the brain stimulates the lateral and posterior groups of hypothalamic nuclei to send nerve impulses via adrenergic sympathetic nerves to cause constriction of cutaneous arterioles and arteriovenous anastomoses so that less blood flows through the skin, the skin temperature falls, and less heat is lost. Exposure to warmth reduces sympathetic activity so that the cutaneous blood vessels dilate passively, more blood flows through the skin, the skin temperature rises, and more heat is lost. Cutaneous vasodilatation is more marked at the extremities, where arteriovenous shunts are most numerous, so the skin temperature of the hands and feet rises until it approaches that of the trunk. At rest, in environmental temperatures in the range 25 to 29°C, core temperature is regulated entirely by vasomotor control (Gagge et al., 1938).

SWEATING. Sweating usually starts when the skin temperature rises to about 34.5°C, which corresponds to an air temperature of about 31°C for a resting nude man or about 29° if he is wearing light clothing (Robinson, 1949). The critical temperature for the onset of sweating is about 2°C higher in women (Hardy, 1961). Sweating occurs at lower environmental temperatures if the rate of heat production is increased by muscular exercise. It depends on stimulation of the sweat glands by cholinergic sympathetic nerves, receiving impulses from the middle group of hypothalamic nuclei, and is initiated by sensory nerve impulses from warmth receptors in the skin, by a rise in temperature of the blood flowing through the brain, and possibly by impulses from mechanoreceptors in muscles and joints (Saltin and Hermansen, 1966).

Sweating of the palms of the hands or the soles of the feet depends on emotional changes rather than on body temperature. Sweating of these regions is not part of the temperature-regulating mechanism, though heat is lost by evaporation of this sweat.

Heat Production

Heat is constantly being produced throughout the body as a by-product of metabolism; at least 95 percent of the energy released by chemical reactions in the body at rest is in the form of heat (Bell et al., 1976). The basal metabolic rate (BMR) for a young adult man at rest in a warm environment at least 12 hours after a meal is about 40 kcal/m² body surface/hr (Ganong, 1977). BMR is highest in young children and falls progressively with advancing age; it is regulated by hormones (especially thyroxine and triiodothyronine). The resting metabolic rate in comfortable, but not basal, conditions is sometimes taken as a unit of energy output (met)* (Bligh and Johnson, 1973). Active exercise can increase the metabolic rate to 10 met or more (Robinson, 1963).

Intake of food increases the metabolic rate. Protein and, to a lesser extent, fat and carbohydrate have a specific dynamic action, stimulating metabolism and so increasing the rate of heat production. Catecholamines, secreted by the adrenal medulla in conditions of stress, also increase the metabolic rate. Neither food nor stress increases the metabolic rate nearly as much as does muscular exercise. Since the mechanical efficiency of skeletal muscle is approximately 25 percent, about 75 percent of the energy expended in muscular exercise is in the form of heat (Åstrand and Rodahl, 1970).

Regulation of Heat Production

The rate of heat production, except under stress, depends mainly on the basal metabolism and the degree of physical activity and is not a function of the temperature-regulating system. If, however, in a cold environment the control of heat loss is inadequate to prevent a progressive fall in core temperature the rate of heat production is increased.

NONSHIVERING THERMOGENESIS. Cooling of the blood, or impulses from cold receptors in the skin, stimulates the posterior and lateral hypothalamic nuclei not only to cause peripheral

---

*1 met = 50 kcal/m² body surface/hr (58 w/m²)

vasoconstriction but also, via sympathetic nerve fibers, to promote the secretion of epinephrine and norepinephrine by the adrenal medulla. Epinephrine increases the metabolic rate of cells throughout the body, and norepinephrine increases the mobilization and oxidation of fatty acids from adipose tissue. Although in many species, and in the human infant, brown adipose tissue is an important source of these free fatty acids, it does not play an important role in adult man.

Another factor in nonshivering thermogenesis is an increase in tone of skeletal muscle; since more muscle fibers are contracting more heat is produced. Nonshivering thermogenesis can increase heat production by as much as 50 percent (Guyton, 1976) and may be sufficient to prevent further fall in core temperature.

SHIVERING. If nonshivering thermogenesis fails to arrest the fall in core temperature, shivering occurs, a coarse tremor of skeletal muscles that can increase the rate of heat production to three times the basal level (Burton and Edholm, 1955). Although it achieves a higher rate of heat production, shivering is less efficient than nonshivering thermogenesis because thermal insulation is reduced by increased blood flow to superficial muscles and convective loss of heat is increased by the movement (Carlson and Hsieh, 1965). In the nude or lightly clad individual at rest shivering usually starts when the air temperature falls below about 20°C, but if the insulating layer of subcutaneous fat is thick, a lower environmental temperature can be tolerated without shivering (Le Blanc, 1954). Shivering is initiated by nerve impulses passing from the posterior hypothalamic nuclei to the muscles concerned.

INDICES OF THERMAL COMFORT

The factors that determine thermal comfort are (Fanger, 1973):

(1) air temperature
(2) mean radiant temperature
(3) relative air velocity
(4) vapor pressure in ambient air

(5) activity level

(6) thermal resistance of clothing

Numerous indices of thermal comfort have been devised, none entirely satisfactory but some of practical use. For temperate climatic conditions the concept of "effective temperature" is appropriate. The stress of a hot environment may be assessed by the "predicted four hour sweat rate" or the simpler "wet bulb globe temperature index." For a cold environment the "windchill index" is useful. Each of these indices is represented by a single figure based on the measurement of several variables.

Effective Temperature

The effective temperature (ET) of an environment (Houghten and Yagloglou, 1923a,b) is the temperature of a still atmosphere saturated with water vapor in which an equivalent sensation of warmth would be experienced. Air temperature is measured with a mercury-in-glass thermometer, relative humidity can be calculated from the difference between wet and dry bulb thermometers (preferably using a whirling psychrometer or an aspirating psychrometer to eliminate the effects of random air movements) and wind velocity is measured with a vane anemometer, thermoanemometer, or katathermometer.* If allowance is made for radiation by using a globe thermometer† in place of the usual dry bulb thermometer, the corrected effective temperature (CET) can be found from globe thermometer and wet bulb thermometer readings and air movement, using a nomogram (Bedford, 1946). There is a "basic" scale for persons stripped to the waist and a "normal" scale (Fig. 2) for those wearing normal indoor clothing. A comfortable CET for sedentary work is in the range 19 to 23°C

---

*The katathermometer is an alcohol thermometer with the bulb coated with highly polished silver to minimize loss of heat by radiation. Since the rate of fall of the alcohol after the instrument has been raised to a temperature above that of the surrounding air is largely dependent on air movement, it can be used to measure this.

†The globe thermometer is a hollow sphere of copper with a black matt surface enclosing a mercury-in-glass thermometer. Like the skin, it freely radiates and absorbs heat; whereas the exposed mercury-in-glass thermometer, acting as a reflector, is relatively insensitive to radiant heat.

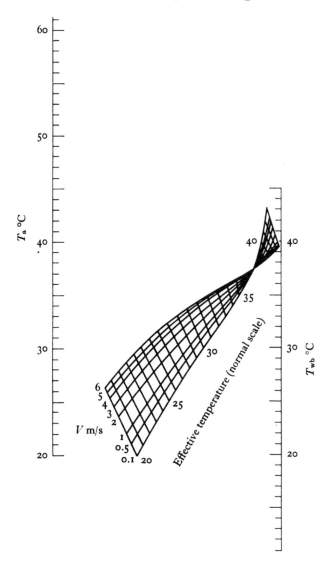

Figure 2. Chart showing normal scale of effective (or corrected effective) temperature. *T*a, dry bulb (or globe thermometer) temperature; *T*wb, wet bulb temperature; *V*, wind velocity (meters/sec). (Reprinted with permission from D. M. Kerslake: *The Stress of Hot Environments.* Cambridge University Press, 1972.)

(Houghten et al., 1929). CET is useful in temperate conditions (Yaglou, 1949) but underestimates the effect of air velocity in a hot, humid environment (Leithead, 1961).

Predicted 4-hour Sweat Rate (P4SR)

The P4SR (McArdle et al., 1947) was designed to assess the degree of heat stress on naval personnel working in tropical conditions. The index, expressed in liters of sweat, is based on air temperature, radiant temperature (often very high in some parts of a ship), humidity, air movement, and rate of working. The basal scale for men at rest, wearing only shorts, is given on a nomogram (Fig. 3). It is modified for men working in overalls by adding 1°C to the observed wet bulb temperature for the clothing and a figure corresponding to the total metabolic rate (shown in the inset) for the work. It is the most complicated of the thermal comfort indices commonly in use but also the most precise (Kerslake, 1972). The upper limit of tolerance is taken as 4.5 liters.

Wet Bulb Globe Temperature (WBGT) Index

The WBGT index (Yaglou and Minard, 1957), much simpler than the P4SR, was designed to measure the unacceptable level of heat stress for military training in the open. The index is based on psychrometer wet bulb and globe thermometer temperature readings:

$$WBGT = 0.7\ T_{WB} + 0.3\ T_{GT}$$

The upper limit of tolerance is about 30°C.

Windchill

Since the rate of heat loss from exposed skin in a cold environment is largely dependent on air movement, the windchill index of Siple and Passel (1945) has been used as a measure of environmental cold stress.

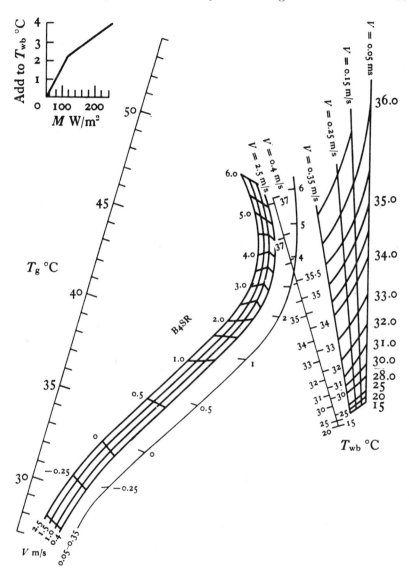

Figure 3. Nomogram for determining predicted four-hour sweat rate (P4SR). B4SR, basal four-hour sweat rate; Tg, globe thermometer temperature; Twb, wet bulb temperature; V, wind velocity (meters/sec). (Reprinted with permission from D. M. Kerslake: *The Stress of Hot Environments.* Cambridge University Press, 1972.)

Windchill (kcal/m²/hr = ($\sqrt{wv \times 100}$ − wv + 10.45) (33 − T_A)

where wv = wind velocity (m/sec)

10.45 = an arbitrary constant

33°C = average skin temperature

T_A = ambient dry bulb temperature (°C).

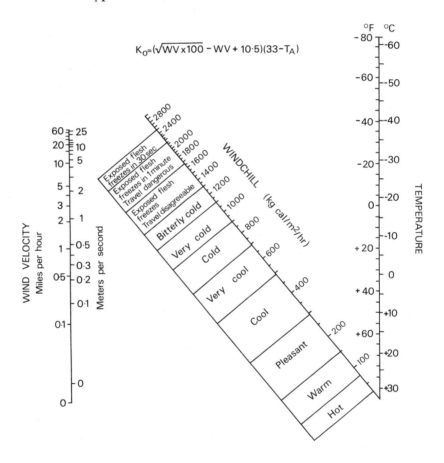

Figure 4. Nomogram for determining windchill from ambient temperature and wind velocity. k_0, windchill; WV, wind velocity; 10.5, arbitrary constant; T_A, dry bulb temperature. (Reprinted with permission from C. F. Consolazio, R. E. Johnson, and L. J. Pecora: *Physiological Measurements of Metabolic Function in Man.* New York, McGraw-Hill, 1963.)

The index may be read off directly from a nomogram or line chart (Fig. 4) if wind velocity and dry bulb temperature are known (Consolazio et al., 1963).

Although the thermal insulation of clothing is a much more important factor than wind velocity for maintaining body temperature in severe cold (Burton and Edholm, 1955), the windchill index has been found useful in practice because tolerance to cold depends largely on cooling of the face and any other exposed part.

CLOTHING

The physiological temperature-regulating mechanism is normally supplemented by wearing appropriate clothing, thin in summer and thick in winter. The main function of clothing in a temperate or cold climate is to increase the thermal insulation of the body by maintaining a layer of still air next to the skin. The "warmth" of a garment depends not on the nature of the material but on the amount of still air trapped between the meshes, air being a very poor conductor of heat (Kerslake, 1972). The fabric should be permeable to water vapor to permit evaporation from the skin and, in a cold climate, the outer layer should be wind-proof. In a hot environment clothing should be loose and porous to permit free evaporation of sweat but thick enough to prevent sunburn. White or light-colored fabrics are preferable to dark because they reflect the sun's rays instead of absorbing them.

Attempts have been made to measure the efficiency of clothing as a heat insulator and various units have been devised. The *clo** unit is convenient for general purposes, 1 clo representing the thermal insulation provided by the normal indoor clothing of a sedentary worker in comfortable indoor surroundings (Bligh and Johnson, 1973). Thermal insulation can be increased to about 2 clo by putting on an overcoat and to as much as 4 clo by special arctic clothing. A limiting factor is the bulk of thick clothing, which adds to the work of body movements and may make fine movements impossible.

---

*1 clo = 0.18°C/kcal/m² body surface/hr
= 0.155°C/m²/W

CHAPTER 2

RESPIRATORY GASES
AND RESPIRATION

ADAPTATION to life at high altitude or in deep water involves adaptation to reduced or increased pressure of the gases in the atmosphere. Gas pressures in the inspired air determine the pressure (tension) of the gases in the blood and other tissues. Physiological adaptation to low gas pressures on a high mountain is more effective than adaptation to high gas pressures under water, which would be an unnatural environment for a terrestrial animal such as man.

RESPIRATORY GASES

Composition of Respired Air

Except at the very high altitudes encountered in space travel (Chapter 8) dry atmospheric air is of very constant composition (20.95% O_2, 79.01% N_2*, 0.04% CO_2). The volume of water vapor

TABLE II

COMPOSITION OF AIR (PERCENT)

Gas	Inspired*	Expired	Alveolar
Oxygen	21	15	13
Carbon dioxide	0	4	5
Nitrogen	78	75	76
Water vapor	1†	6	6

*Figures are rounded off to the nearest whole number.
†Variable but does not affect the relative proportions of other gases.

---

*It is convenient to include with nitrogen, argon (0.93%) and other gases present in much smaller amount (Ne, H_2, He, Kr, Xe).

in the air is variable. Expired air is a mixture of alveolar air, which has lost oxygen to the blood and gained carbon dioxide from the blood, and inspired air from the air passages, which has not undergone any exchanges of gases with the blood (Table II). It is saturated with water vapor that, at body temperature (37°C), exerts a pressure of 47 mmHg.

Gas Pressures

The pressure exerted by each gas in a mixture of gases such as air (partial pressure) is in proportion to its volume (Dalton's law). Hence, in dry air at the mean atmospheric pressure at sea level (760 mm Hg) the partial pressure of oxygen is about 160 mm Hg and of nitrogen about 600 mm Hg. In the alveoli, diluted with carbon dioxide and with water vapor, the partial pressure of oxygen is about 100 mm Hg.

With increase in altitude the atmospheric pressure, and hence the partial pressure of oxygen, falls progressively (Fig. 5). At the increased pressures at which air is supplied to divers (Chapter 5) the gas partial pressures are correspondingly increased.

Gases in Solution

The volume of a gas dissolved in a liquid with which it is in contact depends on the partial pressure of the gas (Henry's law), on the solubility of the gas in the liquid, and on the temperature of the liquid. When equilibrium is established the pressure of the gas in the liquid (tension) is the same as the partial pressure of the gas in contact with the liquid. At a higher temperature the volume of dissolved gas at any given tension is reduced.

RESPIRATION

In its broadest sense respiration includes the intake of oxygen from the environment, carriage of it in the blood to cells throughout the body, utilization of it in the cells with release of energy, and carriage of the resulting carbon dioxide to the

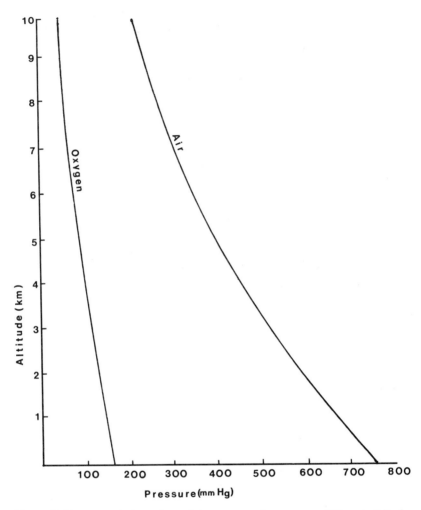

Figure 5. Pressure of air and partial pressure of oxygen at different altitudes. (From data in J. H. Comroe: *Physiology of Respiration*, 2nd ed. Chicago, Year Book Medical Publishers, 1974.)

lungs and thence to the environment. The term is commonly used in its more restricted sense to signify external respiration, the respiratory movements which convey fresh oxygen to the alveoli and remove surplus carbon dioxide and the diffusion of these gases through the walls of pulmonary alveoli and capillaries.

The passage of oxygen from alveolar air to cells throughout the body depends on the pressure gradient from about 100 mm Hg in the alveoli to below 5 mm Hg in active cells (Guyton, 1976). At high altitude the gradient is reduced by the fall in partial pressure of oxygen in the alveoli. The pressure gradient for carbon dioxide is from about 46 mm Hg in the cells to about 40 mm Hg in the alveolar air (Guyton, 1976).

External Respiration

Ventilation

In quiet respiration contraction of the diaphragm and external intercostal muscles increases the volume of the chest, and hence of the lungs (inspiration). Relaxation of these muscles allows the chest and lungs, by their inherent elasticity, to return to their original volume (expiration). Deep respiration involves the action of more muscles both for inspiration and for expiration. Inspiratory movements cannot be performed against a high external pressure, such as is encountered in diving (Chapter 5), unless the gas mixture supplied to the lungs is at a correspondingly high pressure.

Pulmonary Diffusion

Diffusion of oxygen through the thin walls of alveoli and capillaries is so rapid that arterial blood leaving the lungs is normally about 95 percent saturated with oxygen; the difference of 5 to 10 mm Hg in oxygen pressure between alveoli and arterial blood (Heath and Williams, 1977) being mainly due to the passage of a small amount of deoxygenated blood from pulmonary artery directly into the pulmonary veins and to uneven perfusion of the lungs due to the low pulmonary arterial pressure. The increased lung volume and raised pulmonary arterial pressure that occur at high altitude (Chapter 6) increase the pulmonary diffusing capacity for oxygen by 20 to 30 percent (Heath and Williams, 1977) and reduce the alveolar-arterial gradient to about 1 mm Hg (Hurtado, 1963).

Since carbon dioxide diffuses about 20 times faster than oxygen through capillary and alveolar walls (Guyton, 1976) the carbon dioxide tension in arterial blood leaving the lungs is practically the same as the partial pressure of the gas in the alveoli.

Gas Transport in the Blood

Oxygen

The volume of oxygen carried in arterial blood (oxygen capacity) is normally about 19 vols. per 100 vols. of blood (Table III). It depends on the oxygen tension and the amount of hemoglobin in the red cells, which carry most of the oxygen in loose combination with the hemoglobin; but an increase in oxygen tension above 100 mm Hg causes little increase in the volume of oxygen carried, because hemoglobin is already 95 percent saturated with oxygen at this tension. Only the oxygen carried in simple solution in the blood plasma (0.3 ml/100 ml at 100 mm Hg) increases in direct proportion to an increase in oxygen tension. An increase of 20 percent in hemoglobin concentration in the blood increases its oxygen capacity by about 20 percent.

TABLE III

BLOOD GASES
(Normal values at rest BTPS*)

Gas	Arterial		Venous	
	Volume (ml/100 ml)	Tension (mm Hg)	Volume (ml/100 ml)	Tension (mm Hg)
Oxygen	19	100	14	40
Carbon dioxide	52	40	60	45
Nitrogen	1	570	1	570

*Body temperature and standard pressure, saturated (37°C, 760 mm Hg, water vapor pressure 47 mm Hg).

The relation of volume to tension of oxygen in arterial blood may be shown as an oxygen dissociation curve (Fig. 6). The

characteristic sigmoid shape of the curve is due to changes in
the configuration of the hemoglobin molecule and its affinity
for oxygen as it becomes oxygenated. The physiological signifi-
cance of the shape of the curve is that little oxygen is given off
as a result of minor reductions in tension, but at the oxygen
pressures that prevail in the tissues the curve falls steeply. The
curve in venous blood has the same shape but is displaced to
the right (Bohr effect) because the higher tension of carbon
dioxide in venous blood favors the release of oxygen. This
determines the physiological curve, which represents what is
actually happening in the body. Another factor that promotes
the release of oxygen is diphosphoglycerate (DPG) in the erythrocytes, which acts only when bound to hemoglobin and binds
more readily when the hemoglobin loses some oxygen, thereby
accelerating the further release of oxygen. Other factors that
promote release of oxygen from the blood in active tissues are
the acid end-products of metabolism and the rise in temperature that accompanies an increase in metabolism.

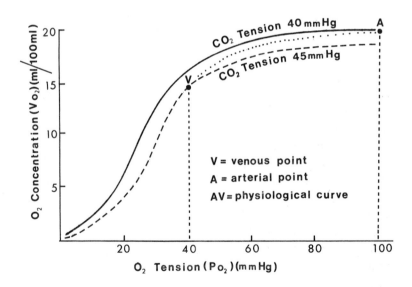

Figure 6. Oxygen dissociation curves of human blood. (Reprinted with
permission from A. W. Sloan: *The Physiological Basis of Physiotherapy.*
London, Baillière Tindall, 1979).

Carbon Dioxide

Carbon dioxide is carried mostly in plasma but partly in the erythrocytes and by far the greater part of it is in chemical combination with constituents of the plasma or the erythrocytes. The rapid combination, which takes place in the venous blood, and the rapid dissociation in the lungs depend on the enzyme carbonic anhydrase in the erythrocytes. Only a small proportion of the carbon dioxide in venous blood is liberated in the lungs (Table III); most of it remains in chemical combination as bicarbonate.

Nitrogen

Nitrogen is carried in simple solution in the blood and is normally in equilibrium with the nitrogen in alveolar air. Being only sparingly soluble in water the volume is low in spite of the high nitrogen tension.

Gas Diffusion in Tissues

The pressure gradient of oxygen from capillary blood to tissues and the volume of oxygen available to the cells are enough to provide all the oxygen required at rest, though actively contracting muscle may outstrip its oxygen supply. Reduced oxygen tension in arterial and capillary blood, such as occurs at high altitude, may reduce the oxygen tension in cells below the critical level (1-3 mm Hg) at which oxidative reactions can take place in the mitochondria (Heath and Williams, 1977). Tissue diffusion may be augmented by increasing the number of open capillaries in the tissue, thereby reducing the distance of the furthest cells from the blood, or by increasing the myoglobin content of the cells, myoglobin having a high affinity for oxygen at low tension. Both these changes occur as a feature of adaptation to high altitude (Chapter 6). There is also an increase in the number of mitochondria in the cell.

REGULATION OF RESPIRATION

Respiratory Centers

The respiratory movements depend on a respiratory center in the medulla oblongata, from which volleys of nerve impulses pass at regular intervals to the muscles concerned with respiration. The activity of the medullary respiratory center is influenced by nerve impulses from higher respiratory centers in the pons, from the cerebral cortex, and from sensory receptors in the medulla oblongata and elsewhere (Fig. 7). It is also sensitive to temperature changes, a rise in core temperature increasing pulmonary ventilation, but this response is not sustained (Chapter 3). The respiratory centers are not clearly defined anatomical entities but merely regions of the reticular forma-

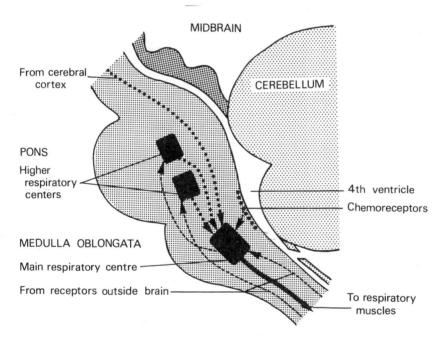

Figure 7. Regulation of respiration. Receptors outside the brain include stretch receptors in lungs, aorta, and carotid sinuses, chemoreceptors in aortic and carotid bodies, muscles and joint receptors, and pain receptors.

tion of the brain stem that are concerned with the regulation of respiration.

Chemoreceptors

The receptors that exert the greatest influence on respiration are the chemoreceptors, which respond to an increase in carbon dioxide tension or to a fall in oxygen tension. The main chemical regulator of respiratory movements is the carbon dioxide tension in arterial blood (Haldane and Priestly, 1905). This is the same as the partial pressure of carbon dioxide in alveolar air (see above) and depends on the rate at which the gas is formed in the body and the rate at which it is expelled from the lungs. The chemoreceptors in the medulla oblongata and in the carotid and aortic bodies respond to a rise in carbon dioxide tension by increasing their stimulation of the respiratory center and to a fall by reducing it. This is a homeostatic mechanism since increased pulmonary ventilation eliminates the surplus carbon dioxide, whereas reduced ventilation allows the concentration in alveoli and blood to rise. The alveolar partial pressure and arterial blood tension of carbon dioxide are normally maintained at about 40 mm Hg, corresponding to 5.3 percent carbon dioxide in alveolar air at an atmospheric pressure of 760 mm Hg (Guyton, 1976). It is the partial pressure of carbon dioxide, not the percentage by volume, that determines pulmonary ventilation; at high atmospheric pressures the proportion of carbon dioxide in the alveolar air is reduced, and at low atmospheric pressures it is increased (Haldane and Priestly, 1905).

The chemoreceptors most sensitive to changes in carbon dioxide tension are situated in the medulla oblongata and respond to a fall in the pH of cerebrospinal fluid due to increased diffusion of carbon dioxide from the blood (Comroe, 1974). Although this is a slow reaction it is very sensitive; an increase of 2 mm Hg in arterial carbon dioxide tension may double the pulmonary ventilation and a fall of the same amount may temporarily arrest it (Haldane and Priestly, 1905); the carotid and aortic bodies respond only to a change of the order of 10 mm Hg (Comroe, 1974).

Oxygen lack (hypoxia) sensitizes the peripheral receptors to carbon dioxide (Asmussen, 1963) and also stimulates them directly, but the direct response to oxygen lack comes into action only when the oxygen tension falls to about 50 mm Hg (Comroe, 1974). Very severe hypoxia, which never occurs under natural conditions, depresses the respiratory center so much that it fails to respond to stimulation and respiratory movements cease.

Voluntary breath-holding is limited by the build-up of carbon dioxide to a breaking-point at a partial pressure of about 48 mm Hg at which it is impossible to restrain respiratory movements (Rahn, 1963).

Other Receptors

The medullary respiratory center is stimulated by nerve impulses from higher centers in the brain in states of excitement or emotion and from muscle and joint receptors during active exercise. It is inhibited by impulses from stretch receptors in the lungs and in the carotid sinuses and aorta (Sloan, 1979). The very considerable increase in pulmonary ventilation during active muscular exercise is not yet fully explained (Comroe, 1974).

CHAPTER 3

HOT ENVIRONMENTS

Extremes of heat are commonly encountered in the tropics and less commonly in temperate zones, where they may be more dangerous due to lack of physiological and behavioral adaptation of the inhabitants to heat. A tropical environment is one in which day-time (and sometimes night-time) temperatures are high, imposing stress on the mechanisms for losing heat.

STRESSES IN TROPICAL ENVIRONMENTS

In a hot environment at rest there is a rise in core and in shell temperature, increased heart rate, increased pulmonary ventilation, and sweating. Peripheral vasodilatation takes place, permitting increased heat loss by radiation, convection and conduction provided the environmental temperature is lower than that of the skin (Chapter 1). In a warm environment a greater proportion of the venous return of blood from the extremities flows in the superficial veins, thereby undergoing more cooling and reducing the countercurrent cooling effect of blood in the deep veins on arterial blood flowing to the extremities (Bazett, 1949); the higher shell temperature of the limbs increases the rate of heat loss. Cutaneous vasodilatation leads to a fall in arterial blood pressure, which may cause faintness and loss of consciousness if the subject stands erect (Robinson, 1972).

With exercise the core temperature is set at a higher level, related to work load but independent, over a wide range, of environmental temperature (Fox, 1974). The increased cardiac output, which results from exercise, counteracts the fall in blood pressure due to vasodilatation, but diversion of blood to the cutaneous blood vessels reduces the supply to active muscles and so reduces the physical work capacity (Wells and Paolone, 1977). The sweat rate is greatly increased over the resting rate at

the same core temperature, which suggests that mechanoreceptors as well as thermoreceptors stimulate the regulating center
(Saltin and Hermansen, 1966; Wyndham, 1973). If the atmosphere is hot and humid the normal loss of heat by evaporation of sweat is reduced, causing a greater rise in body
temperature than in a dry environment at the same temperature. A vicious circle may be established because the rise in core
temperature increases the metabolic rate and the increased heat
production causes a further rise in core temperature (see below).

Active sweating causes dehydration unless a corresponding
amount of water is ingested. The blood becomes more concentrated, and its increased viscosity causes an increased load on
the heart and reduced efficiency of the mechanisms for heat
loss. High osmolality of the blood plasma stimulates secretion
of antidiuretic hormone by hypothalamic nuclei and liberation
of this hormone by the posterior lobe of the pituitary gland
reduces the volume of urine formed, but even in severe dehydration the urine volume does not fall below about 500 ml/day
(Bell et al., 1976). The urine becomes more concentrated since
the excretory products are dissolved in a reduced volume of
water.

A complication of excessive sweating is salt deficiency. Sweat
is a dilute solution (0.2 - 0.5 g/100 ml) of sodium chloride in
water with traces of other substances. Moderate loss of salt in
sweat is compensated by reduced excretion of it in urine, but
excessive sweating depletes the body's salt stores unless salt, as
well as water, is replaced. The onset of salt deficiency is insid-
ious since man, unlike some other animals, has no instinctive
urge to take salt when his body lacks it. Many of the disorders
that used to be considered inevitable in the tropics, including
loss of appetite, muscle cramps, and mental depression and
irritability, are merely manifestations of salt deficiency.

Another danger of a tropical environment is sunburn. Unac-
customed exposure to sunlight may result in severe burning of
the exposed skin. The ultraviolet rays of the solar spectrum,
which are most intense between 10 AM and 2 PM, are the most
harmful in this respect and repeated exposure may lead to
cancer of the skin, especially in people with a fair complexion
(Howell, 1960).

Desert

A hot desert has a mean day temperature of at least 30°C* in the hottest month and annual rainfall less than 30 cm (Lee, 1964). Air temperature is high and solar radiation (direct and reflected from sand) is high because of the dry atmosphere. Since there is little or no shade in the desert there is considerable danger of sunburn.

In the dry heat of the desert loss of heat by evaporation of sweat is usually sufficient to prevent an undue rise in core temperature. Evaporation of water in the lungs also makes a significant contribution to heat loss when the inspired air is hot and dry (Chapter 1). The principal danger is dehydration, all the greater because the supply of water may be limited. If no water is available, survival may be only for a couple of days. Provision of water without salt saves life, but muscle cramps (see below) and other symptoms of salt lack may ensue.

The wide diurnal range of temperature commonly encountered in the desert imposes considerable stress on the temperature-regulating mechanism, since the night, even after a hot day, may be cold. Heat conservation is then the main requirement.

Jungle

The atmosphere of the jungle is usually hot and humid. Little direct sunlight may penetrate to ground level but there is little movement of air, which would aid evaporation from the skin. Profuse sweating occurs but the sweat runs off the skin with little cooling action and little heat is lost by the lungs to the moist inspired air. Usually sufficient water is available to prevent dehydration, but salt deficiency may be experienced. Sunburn is seldom a complication.

The jungle is essentially a hostile environment due to the multitude of predators, parasites, and disease-carrying insects that inhabit it.

---

*Conversion chart for Celsius (centigrade) and Fahrenheit scales of temperature in Appendix.

ADAPTATION TO HEAT

Individual Adaptation

Man probably evolved as a tropical animal and physiological adaptation to heat is more effective than to cold. Acclimatization to heat on moving from a temperate to a tropical environment takes about 2 weeks (Ladell, 1964). In the past this could take place during the journey, but with the development of air travel the move can now be completed in far too short a time for physiological adaptation to take place.

Physiological Adaptation

The core temperature, which rises on entering a tropical environment, falls slowly towards its original level. Heart rate and cardiac output rise with the rise in core temperature and tend to remain elevated when the core temperature falls (Ingram, 1977). Pulmonary ventilation, which is raised at first, returns to its original level (Robinson, 1968). Blood plasma volume increases by as much as 25 percent (Fox, 1965), probably as a result of increased secretion of aldosterone and antidiuretic hormone (Ladell, 1964). There is some evidence of reduced thyroid activity with acclimatization to heat (Samueloff, 1974), and the basal metabolic rate (Chapter 1) may be reduced (Bazett, 1949). Sweating becomes more profuse and starts at a lower environmental temperature (Edholm et al., 1977). The sweat becomes more dilute due to increased aldosterone secretion (Collins, 1963). The capacity to perform prolonged, strenuous physical work in a hot environment is greatly increased by acclimatization (Belding, 1972; Edholm, 1972).

Women, presumably because of their lower B.M.R. or because of a wider range of vasomotor control (Fox, 1977), can tolerate a higher environmental temperature than men at rest but have poorer adaptation to exercise in heat because of their lower sweating capacity (Fox, 1974).

Another important factor in acclimatization is increased tolerance of the discomfort due to heat (Lee, 1964).

Increased pigmentation of the skin gives some protection against sunburn but also increases absorption of solar radiation, raising the shell temperature and promoting sweating.

Behavioral Adaptation

In the tropics active work in the heat of the day should be avoided if possible and full use made of any available shade. Tropical clothing should be light, loose, and porous. In the desert, clothing should be adequate to protect the body from sunburn and from gain of heat by radiation and convection from the environment (Gosselin, 1947).

The diet in the tropics should have a lower energy value than in temperate or cold climates; a reduction of 5 percent has been recommended for every 10°C above an environmental temperature of 20°C (Consolazio, 1963). The diet should be low in protein and in fat because of the specific dynamic action of protein and the high energy value of fat.

Racial Adaptation

Touareg

The Touareg, whose home is the Sahara desert, are a tall and thin people, having a body shape that provides the maximum proportion of surface area for cooling to mass of heat-producing tissue. They wear loose, porous clothing and their behavioral pattern avoids, as far as possible, active exercise in the heat of the day (Beighton, 1971).

Bushmen

The Bushmen of the Kalahari Desert, like the Arabs, show behavioral rather than physiological adaptation to desert conditions (Jenkins and Nurse, 1976). In view of their widespread distribution until a few generations ago, genetic adaptation to the desert is unlikely (Tobias, 1957). They do, however, conserve heat more efficiently at night than Caucasians (Chapter 4).

Negroes

The dark skin of black Africans is a manifestation of racial adaptation to the environment. The pigment (melanin) in the skin protects the skin from sunburn and the underlying tissues from excessive exposure to ultraviolet light and, by absorbing heat rays, raises the skin temperature and promotes sweating. An African's sweat has a lower salt content than that of an unacclimatized European. European and African, when fully acclimatized, start to sweat at the same rectal temperature, sweat at the same rate, and have the same low salt content in the sweat (Ladell, 1964). The range of body dimensions of African tribes, from the very tall Masai to the very small Pygmies, bears no obvious relation to environment.

FAILURE OF HEAT ADAPTATION

Apart from the minor disorder of heat cramps there are two different patterns of failure of adaptation to intense heat: heat exhaustion, characteristic of hot, dry conditions as in the desert, and heatstroke, characteristic of hot, humid conditions as in the jungle. Failure of adaptation may occur during heat-waves in temperate as well as in tropical climates and may involve elements of both these conditions. It is commoner in the physically unfit and in the elderly (Wyndham, 1973).

Heat Cramps

Heat cramps are a complication of salt deficiency, usually due to excessive sweating with replacement of water but not of salt. The osmolality of the extracellular fluid is reduced and water passes into the cells by osmosis. When the cells of skeletal muscle take up water in this way they become stiff and painful (cramp). This is most likely to occur in the most active muscles, where the end products of metabolism increase the osmolality of the intracellular fluid.

The incidence of heat cramps is highest in people performing hard physical work in a hot, humid environment, e.g. miners' cramps, stokers' cramps. For prevention or treatment thirst should be assuaged by drinking a dilute saline solution (about 1%) or salt tablets may be swallowed along with water or extra salt added to food.

Heat Exhaustion

In a hot, dry environment free evaporation of sweat may be sufficient to prevent an undue rise in body temperature during the heat of the day but there is considerable danger of dehydration due to loss of water in sweat and from the lungs. Vasodilatation in the skin, tachycardia, and the reduced volume and increased viscosity of the blood may lead to peripheral circulatory failure. Loss of water equivalent to 15 percent of the body weight is incapacitating and loss of about 20 percent is fatal (Brown, 1947). Thirst is an inadequate indication of the degree of dehydration, seldom promoting replacement of more than two-thirds of the water lost (Robinson, 1968). Provision of water, without salt, is life-saving, but muscle cramps and other symptoms of salt lack may ensue.

Heatstroke

Heatstroke during strenuous physical exercise is notorious in oil tanker crews in the Persian Gulf, black miners in South Africa, and army recruits training in a hot environment (Shibolet et al., 1976). It occurs also in urban areas in the United States during heat waves.

In a hot, humid environment, where sweating is profuse but evaporation limited, the core temperature tends to rise progressively until the temperature-regulating mechanism breaks down and sweating ceases. Further rise in temperature causes headache, irritability, or even delirium (Manson-Bahr, 1966), and there may be irreversible damage to brain cells. A core temperature about 40°C is dangerous and the critical temperature for death from heatstroke lies between 42 and 45°C, depending on the length of time that the individual is exposed to it (Lee, 1964).

Although loss of water may be considerable, heat exhaustion is uncommon in the jungle because water is usually available for replacement, but a combination of heatstroke and heat exhaustion may occur.

The incidence of heatstroke is low at effective temperatures below 25.5°C and high at effective temperatures above 33.3°C (Wyndham, 1970). The danger is reduced by avoiding physical work, as far as possible, in a hot, humid environment; and by wearing light, loose clothing, which permits such evaporation of sweat as the environment permits. More heat is lost when sweat evaporates directly from the skin than when it must first penetrate clothing (Kerslake, 1972). In American military training establishments the incidence of heat casualties was drastically reduced when active outdoor training of new recruits was stopped when the WBGT index (Chapter 1) reached 29.4°C; after three weeks of training this could be raised to 31.0°C (Yaglou and Minard, 1957).

Heatstroke calls for urgent treatment, directed to lowering the core temperature. Ideally the patient should be packed in ice but, failing this, he may be covered with a light cloth, which is kept moist and fanned to promote evaporation. Since the fall in temperature, once it starts, tends to progress, the treatment should be stopped when the patient's rectal temperature falls to 39°C (Maegraith, 1971).

DEEP MINES

Mines are an example of an artificial tropical environment being, in most cases, much deeper than natural caves. As a general rule the deeper the mine the higher the temperature, and most mines are humid. Some of the gold mines near Johannesburg are more than 3000 m deep and the rock must be sprayed with water to reduce the incidence of dust diseases of the lungs. Since mining is heavy work, tending to raise core temperature, and the humidity reduces evaporation of sweat, there is considerable danger of heatstroke. Although economic considerations make air conditioning of mines impracticable, forced ventilation is essential for tolerable working conditions.

The wind velocity achieved by ventilation in gold mines is about 0.4 m/sec and the air is saturated with water vapor (Wyndham et al., 1973). Increasing the wind velocity up to 2 m/sec increases working efficiency (Strydom et al., 1963) but is seldom practicable; further increase confers little benefit.

The incidence of heatstroke in South African gold mines has been drastically reduced by acclimatization and selection of mine recruits. The routine, adopted after numerous experimental trials, is for the recruits to work at least 4 hours a day in a chamber with air at 31.7°C, saturated with water vapor and moving at about 0.4 m/sec (Strydom and Kok, 1970; Wyndham, 1973). The work level is increased progressively over a period of 8-9 days. Oral temperature is measured hourly and the man stops work if it reaches 39°C. The 3-4 percent of recruits who fail to adapt are drafted to other work. Black mine workers usually show some initial acclimatization to moist heat but the same maximum degree of acclimatization is reached by white as by black miners (Lee, 1964). Maximum acclimatization is achieved only by strenuous work in the hot environment (Strydom et al., 1966). It declines progressively over three weeks when the miner is moved to a cooler environment (Williams et al., 1967).

Recently evidence has been adduced that trainee miners adapt more rapidly to heat on a daily dose of 250 mg ascorbic acid (Strydom et al., 1976); it is suggested that the stress of hard work in a hot, humid environment may promote subclinical ascorbic acid deficiency. For work in the worst conditions the miner may wear a climatic suit, comprising a prefrozen waistcoat and thermally insulating over-jacket (Strydom et al., 1975).

CHAPTER 4

COLD ENVIRONMENTS

EXTREMES of cold may be encountered in temperate as well as frigid zones, especially at high altitude; but they are characteristic of the Arctic and Antarctic, both of which include regions of high altitude. The Greenland ice-cap rises to 3,000 m and the Central Antarctic Plateau is 3,200 m above sea level.

STRESSES IN POLAR* ENVIRONMENTS

The principal stress imposed by a polar environment is extreme cold. The mean annual temperature at the South Pole is -50°C† and it falls as low as -82°C in winter (Brooks et al., 1973). Other stresses encountered on polar expeditions include social deprivation and isolation, disruptions of circadian rhythms, and, at high altitude, hypoxia. Due to the snow-covered terrain, high winds, and the "hobbling" effect of polar clothing the energy expended on outdoor activities is about twice as much in a polar as in a temperate environment (Brotherhood, 1973; Tikhomirov, 1973). The dangerous terrain is responsible for many accidents, especially among inexperienced personnel, and disorders of the skin are common due to bulky clothing and limited washing facilities (Lloyd, 1973). Unless sun goggles are worn, there is danger of snow blindness, damage to the retina of the eye by the high intensity of sunlight reflected from snow.

PROTECTION AGAINST COLD

Man evolved as a tropical animal with a narrow range of tolerance of cold (Scholander et al., 1950; Rogers, 1973). He

---

*It is convenient to use the term "polar" to refer to both arctic and antarctic regions.
†Conversion chart for Celsius (centigrade) and Fahrenheit scales of temperature in Appendix.

lacks the thick fur or thick layer of subcutaneous fat which provides thermal insulation for polar animals. Clothing, shelter and heating are essential for his survival in a very cold environment.

Clothing

The fur garments worn by Eskimos are suitable for a polar climate because they are windproof, permeable to water vapor, and hold still air between the hairs; the material of which polar clothing is made is of little consequence provided it fulfills these criteria.

The polar clothing, which proved satisfactory in the British North Greenland Expedition of 1952-54, has been fully described by Lewis and Masterton (1955). It comprised a string vest, long underpants, a flannel shirt, windproof combat jacket and trousers, parka (a windproof jacket lined with wool, with cowled head and long tail which can be brought between the legs to button in front), one or two pairs of gloves and two of mittens, three or four pairs of socks, and leather boots with a thick plastic insole. In very cold weather canvas mukluks were worn instead of leather boots to allow sweat to evaporate instead of freezing inside the boot. Outer clothing was fitted with zip fasteners for easy opening or removal during strenuous activity. Underpants were slit up the back to permit defecation without removing them. Sun goggles were worn. Similar clothing has been worn on more recent polar expeditions (Fig. 8).

The thermal insulation of polar clothing can be derived roughly from a formula, giving outer windproof garments (anorak and trousers) and long underpants a value of two layers each and all other garments a value of one layer (Rogers and Sutherland, 1974).

$$\text{Thermal insulation (clo)} = 0.08 \times \text{total number of layers} + 0.51$$

Shelter and Heating

In the igloo, in which Eskimos traditionally spend the

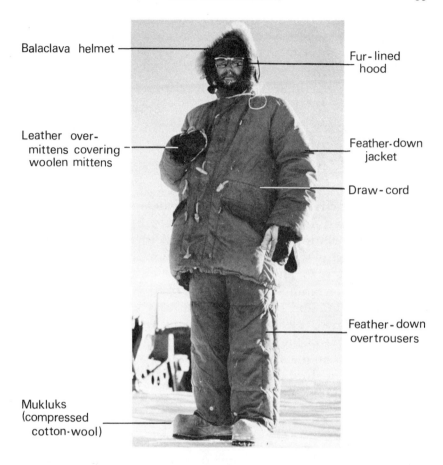

Balaclava helmet

Fur-lined hood

Leather over-mittens covering woolen mittens

Feather-down jacket

Draw-cord

Feather-down overtrousers

Mukluks (compressed cotton-wool)

Figure 8. Clothing worn on Antarctic expedition. (Courtesy of Dr. J. G. McPherson). Underclothing comprised cotton underpants, string vest and quilted bootees, and woolen long johns, vest, tights, shirt, trousers, jersey, socks (2 pairs) and mittens. Long leather overmittens were worn for sledding.

winter, air temperature is about 20°C during the day and about 10°C at night (Weiner, 1964). Tents or huts in polar regions require, like the igloo, effective thermal insulation and controlled ventilation. Huts are usually prefabricated of insulating material, e.g. glass wool between plywood, windows are triple glazed, and doors are double. Heating may be by oil stoves or by electric radiators, run by a generator which also supplies power for light and for instruments if the purpose of the expedition is scientific (Lewis and Masterton, 1955). Since ventilation is a problem, particularly if the tents or huts are buried in snow, there is a danger of carbon monoxide poisoning from slow-burning stoves.

In a tent, where less wall insulation is possible, a good sleeping bag is essential; this can have thermal insulation up to 11 clo (Burton and Edholm, 1955).

Food

Since the energy expended on physical work in a polar environment is high, and the metabolic rate may be raised during periods of inactivity to maintain body temperature, a high-energy diet is required. The Food and Agriculture Organization of the United Nations recommends an addition of 3 percent of the normal energy value for every 10°C below an environmental temperature of 10°C, and the National Research Council of the United States recommends an increase of 5 percent for the first 10°C below 20°C and of 3 percent for each subsequent fall of 10°C (Consolazio, 1963). The individual energy requirement depends largely on the degree of activity, ranging from about 3,600 kcal/day at the base for sedentary work to about 5,500 kcal/day during sledding (Edholm and Lewis, 1964). The diet should have a high content of protein because of its specific dynamic action, and of fat, which is the richest source of energy.

If the group is to be isolated for a prolonged period, careful planning of the diet is essential to ensure that it will be adequate in energy value and not lacking in vitamins or other essential constituents. Facilities must also be provided to supply water by melting snow or ice.

Medical Care

An isolated group must have adequate medical supplies, which have been listed by Lewis and Masterton (1955). If practicable, the group should include a medical practitioner. Dental treatment is commonly required, but makeshift dentistry can be performed by a doctor with some preliminary training in dental conservation and extraction.

ADAPTATION TO COLD

Maintenance of Body Temperature

Although man cannot survive in a polar environment without artificial aids, in less extreme cold the temperature-regulating mechanism may be sufficient to maintain body temperature. Reduction of heat loss by peripheral vasoconstriction can maintain a normal core temperature in a lightly clad individual at ambient temperatures down to about 25°C (Gagge et al., 1938).

Countercurrent heat exchange between the limb arteries and their accompanying veins reduces the loss of heat from hands and feet in cold weather. Transfer of heat from arteries to veins reduces the skin temperature at the extremities, so less heat is lost from the body; and transfer of heat warms the cooled venous blood, so there is less fall in core temperature (Bazett, 1949; Thompson, 1977).

At temperatures about 25°C nonshivering thermogenesis occurs (Chapter 1), and shivering thermogenesis begins below about 20°C. The increased metabolic rate in the cold is brought about by increased secretion of the hormones of the adrenal medulla (Bodey, 1973) and of the thyroid gland (Leith, 1973). Epinephrine also raises the arterial blood pressure, which inhibits secretion of antidiuretic hormone and so promotes diuresis (Suzuki, 1972).

Individual Adaptation

In acclimatization to a cold environment the more efficient nonshivering thermogenesis takes the place of shivering, the adaptation taking about six weeks (Gelineo, 1964). Less heat is produced on exposure to cold in the acclimatized, their core temperature being allowed to fall to a lower level (Le Blanc, 1956). Further evidence of acclimatization is that members of polar expeditions wear less clothing after some months of exposure to cold (Goldsmith, 1959; Budd, 1973).

Physically fit individuals have better tolerance of cold than the less fit, possibly because of more active peripheral circulation (Le Blanc, 1975). Adaptation may be local. During winter, fishermen in the Gulf of St. Lawrence immerse their hands for many hours in water below 10°C without discomfort. They have increased local blood flow, but show no general adaptation to cold (Le Blanc, 1975).

Racial Adaptation

In populations that have inhabited a cold climate for many generations a degree of natural selection is to be expected. The limits of such racial adaptation are set by the genetic capacity. Natural selection operates on the capacity for physiological adaptation, not directly on the adaptation (Prosser, 1964).

Eskimos

The Eskimos, who inhabit arctic regions, are a short, broad people, a body shape which allows the maximum of heat-producing tissue for the minimum area of cooling surface. Their subcutaneous fat, though adequate for thermal insulation, is less than that of White Americans (Hildes, 1963). Their traditional fur clothing has high thermal insulation. With adaptation to a European pattern of life the differences between Eskimo and Canadian body build are disappearing (Godin and Shephard, 1973). Contrary to previous reports, Godin and Shephard found no increase in the metabolic rate of Eskimos over that of Caucasians either under resting conditions or when performing the same activity.

Ainu

The Ainu are a primitive people inhabiting Hokkaido, the northern island of Japan, where the winter climate is very cold. Their physiological responses to cold do not differ significantly from those of other Japanese living in Hakkaido, but the concentration of free fatty acids (FFA) in the bloodstream during fasting is lower in the Ainu, and their basal metabolic rate is inversely related to FFA concentration, whereas the relationship is positive in other Japanese (Itoh, 1974). It is assumed that this is due to a more active turnover of FFA in the Ainu.

Alacaluf Indians

The Alacaluf Indians of Tierra del Fuego wear clothing which, by European standards, is quite inadequate for their very cold environment. Their resting metabolic rate is high (about 60 kcal/m²/hr), and on a very cold night it falls to the level to which the metabolic rate of a Caucasian would rise under these conditions (Hammel, 1964). The mean skin temperature falls below that of a Caucasian, which reduces the rate of heat loss, while the rectal temperatures are about the same. The foot temperature is higher in the Alacaluf Indian, which is possibly a protective mechanism against frostbite.

Australian Aborigines

Aborigines are small people with little subcutaneous fat. They wear no protective clothing against cold exposure though the temperature in Central Australia may fall to 0°C. They show some behavioral adaptation, sleeping between campfires often protected by primitive windbreaks (Le Blanc, 1975). During sleep their shell and core temperatures fall below the level at which a Caucasian would start to shiver (Scholander et al., 1958). The reduction in temperature gradient between skin and environment reduces the rate of heat loss.

Kalahari Bushmen

The Bushmen are small people of a mean height of 145 to 150 cm (Harrison et al., 1964). Their mean body fat (7.7%) is the lowest of any racial group studied (Hammel, 1964) except Sherpas (Sloan and Masali, 1978). The Kalahari desert that the bushmen inhabit has a climate similar to that of Central Australia. Bushmen wear very little clothing — a loincloth of skin for the men and a kaross of skin for the women (Hammel, 1964). They sleep under skins beside a fire and, like the Aborigines, may construct primitive windbreaks (Wyndham and

Morrison, 1958). During a cold night their skin temperature falls more than that of lightly clothed Caucasians exposed to the same conditions, but the rectal temperature is about the same in the two groups (Ward et al., 1960). The degree of peripheral cooling that reduces heat loss in the Bushmen, who are never exposed to extreme cold, would be dangerous in the Eskimos.

FAILURE OF COLD ADAPTATION

Prolonged exposure to cold, with inadequate clothing, may cause a fall in core temperature to below 34°C (hypothermia) or local damage to the extremities (frostbite or immersion foot). These conditions may occur during winter in normally temperate climates, especially in the elderly and ill-nourished.

Hypothermia

If the temperature-regulating mechanism is inadequate to maintain body temperature in a very cold environment the core temperature falls progressively. Below about 34°C the individual becomes drowsy and his reactions sluggish; below about 30°C consciousness is lost. At about 28°C the action of the heart becomes irregular (usually atrial fibrillation) and at about 25°C death occurs, usually from ventricular fibrillation (Hock and Covino, 1958).

Metabolic acidosis may occur in hypothermia due to failure to metabolise lactic acid, depressed respiration, and increased solubility of carbon dioxide at low temperatures (Keatinge, 1969).

Treatment of Hypothermia

The treatment of hypothermia is to rewarm the patient, but the rewarming must either be very rapid or slow (Herrington, 1949). Moderate rewarming can be lethal, because it transfers cold blood from the periphery to the deep vessels when the peripheral vessels open up, so causing a further fall in core temperature and possibly leading to ventricular fibrillation.

Recovery has occurred, with slow rewarming, from a rectal temperature of 18°C (Burton and Edholm, 1955). To counteract severe acidosis sodium bicarbonate may be administered intravenously.

Artificial Hypothermia

Artificial hypothermia is employed in "open heart" surgery to increase the time during which the brain can be deprived of blood (by reducing its metabolic rate) and to facilitate surgical maneuvers by slowing or stopping the heartbeat. The early practice was to anesthetize the patient, administer drugs to prevent shivering (Hock and Covino, 1958), and immerse him in ice. When the core temperature was reduced to 30°C the surgeon could operate directly on the heart, depriving the brain of blood for up to about 14 minutes without ill effects. This "smash-and-grab" technique was later replaced by external cooling, the blood being passed through a cooling coil. For long cardiac operations the heart is by-passed and the circulation maintained by a mechanical pump; the external circuit includes a cooling coil and a gas-exchange system. With the core temperature maintained at 16 to 20°C the surgeon can operate, without restriction of time, on a motionless, bloodless heart; the surgeon can even transplant a heart from another individual (Barnard, 1967).

Usually the heart begins to beat again spontaneously when the patient is gradually rewarmed after the operation, but an electric shock to the heart may be necessary to initiate normal rhythm or to convert fibrillation to a regular rhythm.

Frostbite

Exposure to cold may cause local damage (frostbite). Vasoconstriction reduces the cutaneous circulation in the extremities (toes, fingers, nose, and ears), which are normally at a lower temperature than the rest of the body, below the minimum necessary to maintain the tissues. Countercurrent heat exchange between the limb arteries and their accompanying veins, which serves to maintain core temperature by reducing heat loss, is another factor in reducing shell temperature at the periphery of a cold limb (Bazett, 1949)

The sequence of events in frostbite is intense local vasoconstriction, followed by vasodilatation with increased permeability of the walls of blood vessels, loss of fluid into the tissues, and subsequent hemoconcentration with increased blood viscosity. This "cold vasodilatation" may be due to paralysis of vasoconstrictor nerves (Spealman, 1968) and may at first be intermittent, the local warming resulting from vasodilatation temporarily relieving the paralysis (Burton and Edholm, 1955). Eventually when the shell temperature falls below 2.5°C ice crystals form in the tissues; although most of these are extracellular, the resulting osmotic withdrawal of water from the cells raises the intracellular electrolyte concentration to a level which interferes with cell function (Keatinge, 1969).

Frostnip

A minor degree of frostbite (frostnip) involves only local vasoconstriction without freezing and permanent damage to tissues.

Treatment

The treatment of frostnip is local warmth but the treatment of frostbite is warming of the whole body rather than of the damaged parts, which must be protected from overheating or pressure because of the danger of gangrene. If gangrene occurs, amputation may occur spontaneously, or surgical amputation may be necessary to prevent the spread of infection.

Immersion Foot

Immersion foot is due to prolonged exposure of the feet to cold and wet. Wet skin cools faster than dry and freezes without the supercooling that is necessary for freezing of dry skin (Wilson, 1973). Immersion foot was common during the trench warfare of World War I ("trench foot"). The foot is at first cold, pale, and swollen and later becomes hot, red, and painful due to cold vasodilatation. If untreated, blisters appear and gangrene may set in. Nerve and muscle degeneration is common (Burton and Edholm, 1975). Prevention involves provision of dry socks and boots for soldiers to change into after a spell of duty, and treatment is as for frostbite. A chilblain, due to prolonged local and general cooling, resembles mild immersion foot.

SHIPWRECK

The chief danger of immersion in the sea, apart from drowning, is hypothermia (Keatinge, 1969). The rate of heat loss in cold water is about twenty times greater than in cold air because of the greater specific heat of water and loss of the insulating layer of air when the clothes are saturated with water (Molnar, 1946).

The time of survival in the sea after shipwreck is directly related to the temperature of the water. On immersion in water below body temperature the shell temperature and core temperature fall. In water warmer than 23°C increased heat production can keep pace with the rate of heat loss, which is reduced by the fall in body temperature and the consequently reduced temperature gradient between skin and environment (Hayward et al., 1975); at lower temperatures the cooling is progressive (Molnar, 1946). Most men rescued within 1 hour from water at 5 to 10°C survive but suffer from immersion foot and hand. Frostbite does not occur because sea water is never colder than -1.9°C. Obesity, by increasing thermal insulation, reduces the rate of heat loss (Pugh et al., 1960; Keatinge, 1969).

There is no general agreement whether a shipwrecked man should merely hang on to some floating object or should increase his rate of heat production (and heat loss from convection in water) by swimming for as long as he can. Glaser (1950) recommended swimming for as long as possible, but, unless one is wearing a life jacket and near the shore, the more passive course is preferable (Hayward et al., 1975; Cooper, 1976).

The chance of survival in small boats or rafts after shipwreck in cold water is better if the individual has not been immersed in the sea since dry clothing is a much better thermal insulator than wet. The commonest cause of death at ambient temperatures below 5°C is hypothermia. Other dangers include injury, dehydration, and starvation. Drinking sea water, which is hypertonic and enhances the effects of dehydration, reduces the expectation of life (McCance et al., 1956).

CHAPTER 5

DEEP WATER

MAN is a terrestrial animal and cannot survive more than a few minutes under water unless provided with oxygen, either in air or in an artificial mixture of gases. In the ocean he is liable to be attacked by marine animals, especially in tropical and subtropical seas; these marine animals include sharks, barracuda, coelenterates, echinoderms, and venomous fishes (Halstead, 1976). In deep water man is subjected to great external pressure, to cold, and to darkness.

Without special equipment, diving is limited to a depth of about 25 m (80 ft) and is seldom deeper than about 5 m. With breathing apparatus a diver can descend to much greater depths (Fig. 9). In recent years the search for oil on the continental shelf and the need to recover sophisticated military equipment

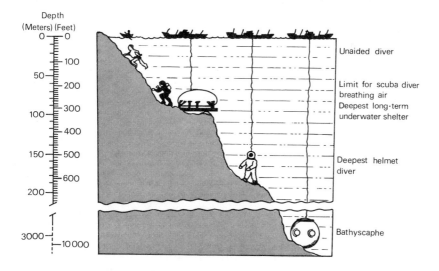

Figure 9. Underwater depths reached with different types of diving equipment. (Reprinted with permission from A. W. Sloan: *Physiology for Students and Teachers of Physical Education.* London, Arnold, 1969.)

from the floor of the ocean have been responsible for very active development of techniques for deep-sea diving. The open sea depth record for helmet diving, established by the U. S. Navy in the Gulf of Mexico in 1975 was 350 m (Kindwall, 1976).

STRESSES IN DEEP DIVING

Pressure

The external pressure on a diver increases by approximately 1 atmosphere (760 mm Hg) for every 10 m (33 ft) of depth in sea water. It is usually expressed in ata units (atmospheres absolute), being the pressure at the surface (1 atmosphere) plus 1 atmosphere for every 10 m of depth. This pressure, even at a depth of 10 m (2 ata) makes breathing impossible unless air or an artificial gas mixture is inhaled at a similar pressure so that the chest can expand on inspiration.

According to Boyle's law the volume of a gas varies inversely with the pressure to which it is subjected. When the external pressure is doubled, the air that has been taken in at the surface is compressed to half its original volume, and air taken in at a depth of 10 m from breathing apparatus at a pressure of 2 ata will expand to twice the normal lung volume on ascent to the surface. Since the pressure increase with increasing depth is linear, descent to 20 m will compress the lung to one-third and to 30 m to one-quarter of its original volume; the greatest volume change therefore takes place near the surface and a pressure change of 1 ata deep in the ocean makes little difference to body gas volumes.

Solution of Gases

The high pressure of gases in the body in deep diving raises the concentration of gases in solution in the body fluids (Chapter 2). The nervous system, which is rich in lipids, preferentially absorbs nitrogen, which is readily soluble in lipids. The time to saturation is different for different tissues but full saturation with nitrogen takes more than 12 hours on a deep dive (Hempleman, 1975).

Gas Intoxication

The high pressure at which gases must be supplied in deep diving introduces an additional hazard since both oxygen and nitrogen at high pressure have toxic effects (see below).

Barotrauma

The rapid pressure changes imposed by rapid descent or ascent are particularly dangerous because of excessive compression of the lung on descent and too rapid expansion on ascent (see below). The eardrum also is liable to damage if the pressure in the middle ear differs greatly from the external pressure.

Cold

Deep diving exposes the diver to cold; the sea temperature tends to become progressively colder with increasing depth (except in polar regions) down to about 1000 m, where it is close to 0°C (Webb, 1975). Because of the greater thermal conductivity of water, loss of heat from the body is more rapid in water than in air.

Darkness

Sunlight penetrates only to a limited depth even in clear sea water, the spectrum being cut off from the red end. The prevailing color underwater is blue-green and at depths below 1000 m there is total darkness.

ADAPTATION TO DIVING STRESSES

There is no evidence of genetic adaptation to the underwater environment, even in the Ama women of South Korea and Japan, who have been practicing breath-hold diving as an occupation for centuries (see below). There are, however, physiological differences between the diving women and others of the same population. The diving women have a larger vital capacity (Kobayasi, 1977), probably from greater development of the inspiratory muscles (Hong and Rahn, 1976); their basal metabolic rate increases by about 35 percent in winter, whereas that of the women not exposed to cold water at this time does not increase (Hong, 1963); and the diving women can stand lower water temperatures without shivering (Rennie et al., 1962).

Experienced divers can tolerate higher pressures of carbon dioxide and lower pressures of oxygen (Schaeffer, 1975; Anthonisen, 1976) and are less susceptible to the complication of rapid pressure changes (Walder, 1975); these adaptations are probably behavioral.

The Diving Reflex

An adaptation to the underwater environment, which man shares with diving mammals, is the diving reflex. Immersion of the face in water, even with the rest of the body exposed to air, causes reflex slowing of the heart and peripheral vasoconstriction (Andersen, 1966). The bradycardia reduces the rate of oxygen consumption by the heart and the vasoconstriction reduces tissue blood flow and oxygen consumption and maintains the arterial blood pressure (Landsberg, 1975). In a prolonged breath-hold dive cardiac arrhythmias are common (Scholander, 1964; Hong, 1976).

DIVING EQUIPMENT

Breath-hold Diving

The simplest form of diving, breath-hold diving, requires no special equipment except a face mask (see below). The duration of the dive is limited by the accumulation of carbon dioxide in the body which, when it reaches the "breaking point," makes breathing imperative (Chapter 2). The dive can be prolonged by preliminary hyperventilation, but this is dangerous for inexpert divers, because although it eliminates some carbon dioxide, it does not significantly increase the store of oxygen; consequently hypoxia may develop before the "breaking point" for breathing is reached, and the diver may lose consciousness. Breath-hold dives usually last about 30 seconds and the maximum duration is about 2 minutes.

The dangers of barotrauma to lungs or ear in breath-hold diving are minimal, because external and internal pressures both increase with depth; but wearing eye goggles is dangerous because the air pressure in them does not increase, and the rise in body pressure (including blood pressure) on descent may cause hemorrhage from the conjunctiva (Hong and Rahn, 1976). This can be prevented by wearing a face mask over eyes and nose so that the pressure in the mask is that in the respiratory passages.

The supply of oxygen to the blood is well maintained during a breath-hold dive, because the rise in alveolar pressure with descent compensates for the oxygen used up and maintains the pressure gradient for oxygen. The carbon dioxide gradient from blood to alveoli falls and may even be reversed, with carbon dioxide passing from alveolar air to blood (Hong, 1976). The dangerous period is the ascent, when the oxygen tension may fall to a critically low level.

Ama

The Ama usually dive to about 5 m, but can attain a depth of about 20 m. Their traditional garb is light cotton clothing, but today many wear a wet suit (see below). Just before diving they hyperventilate moderately, and at the last breath, fill their lungs with air to about 85 percent of their vital capacity (Hong et al., 1963).

Shallow diving is usually "unassisted," with the dive lasting about 30 seconds and repeated after 30 seconds on the surface, during which the catch is transferred to a net suspended from a float. For deeper dives the diver is "assisted" by someone in a boat on the surface, who winches her up with a rope and pulley; this permits a dive of about 60 seconds, but the recovery time at the surface is about 60 seconds so the total time underwater is about the same as for unassisted diving (Hong et al., 1963).

Cold is a limiting factor for the Ama. During 1 hour of

repetitive diving in traditional costume the oral temperature drops to about 35°C in summer and to about 33°C in winter, after which the Ama warm up at open fires on the shore (Hong et al., 1963).

The only special equipment worn by the Ama is a face mask, essential for clear vision. Visual acuity in water without a mask is greatly reduced, because there is little difference in refractive index between water and the cornea. With a mask it is restored to normal, though objects appear nearer. To avoid damage to the eyes by the pressure change during descent the mask must cover eyes and nose (see above).

Diving Bell

If oxygen is supplied to the diver the duration of the dive can be greatly increased. The diving bell, an air-filled chamber open at the bottom, has been in use since the sixteenth century (Bachrach, 1975) but was of limited value until pumps were available to supply the bell with air under pressure from the surface. The caisson (used in building the foundations of docks and bridges) is such a pressure chamber; it is provided with air locks for passing material into and out of the chamber and a decompression chamber, in which the men coming off duty can be slowly reduced to atmospheric pressure, thereby avoiding the dangers of decompression sickness (see below).

Helmet Diving

Although a diving bell provides air and protection to the diver it confines him to a very limited area. Providing him with a helmet, to which air is supplied under pressure by a pump on the surface, allows more freedom of movement and makes possible descent to greater depths. The modern helmet diver wears a watertight "dry suit" over several layers of thick underclothing for thermal insulation.

Limiting factors in deep helmet diving include the resistance of dense compressed gas to flow along a narrow tube and the toxicity of both oxygen and nitrogen at high pressure (see below). For deep diving a gas mixture is used in which nitrogen is replaced by helium, which is less toxic, and the proportion of oxygen is reduced so that its partial pressure does not exceed 0.5 atm (Miller, 1975). If expired gases pass directly into the surrounding water there is no undue accumulation of carbon dioxide but, if there is recirculation of gases to conserve helium, a carbon dioxide absorber must be included in the circuit.

Scuba Diving

The development of SCUBA (self-contained underwater breathing apparatus) from the aqualung invented by Cousteau and Gagnan in 1943 (Cousteau, 1953) further increased the diver's mobility. The gas supply is carried to a face-mask from cylinders on the back and, in most patterns, expired air is breathed out through a valve into the surrounding water. An important feature is the demand valve between cylinder and face-mask, which admits gas, at the pressure of the surrounding water, only during inspiration (Fig. 10).

Gas Mixtures

For shallow dives compressed air is used but at depths greater than 30 m (4 ata) nitrogen is toxic and at depths in excess of 90 m (10 ata, partial pressure of oxygen 2 atm) oxygen is toxic (see below). For deep dives a mixture of oxygen and helium is used. Helium under pressure is less toxic than nitrogen and is also less dense, offering less resistance to flow. Its disadvantage is a high thermal conductivity, which increases the rate of heat loss. Recently a mixture of hydrogen and oxygen has been used, offering even less resistance to gas flow; this is safe from combustion provided the proportion of oxygen in the mixture is less than 6.1 percent (Dougherty, 1976).

Rebreathing

Since provision of gas at high pressure rapidly exhausts the supply a closed circuit apparatus is desirable for deep dives. The simplest such device comprises a tank of pure oxygen, a

Figure 10. Skin-diver with self-contained underwater breathing apparatus (SCUBA). (Reprinted with permission from A. W. Sloan: *Physiology for Students and Teachers of Physical Education.* London, Arnold, 1969.)

bellows into which the diver can breathe to and fro, a canister of baralyme or soda lime to absorb carbon dioxide (Miller, 1975), and a valve to regulate the supply of oxygen. Owing to the danger of oxygen poisoning some dilution of oxygen is required for deep dives. This demands a sophisticated device to monitor the pressure of oxygen in the face-mask; such equipment in turn demands frequent maintenance and adequate training of the diver in its use (Egstrom, 1976). The pressure of oxygen should be between 150 and 400 mm Hg and that of carbon dioxide should not exceed 7 mm Hg (MacInnis, 1966).

Wet Suit

Although a "dry suit," such as that of a helmet diver, may be worn for SCUBA diving, the difficulties of keeping it watertight are such that a "wet suit" is usually preferred (Fig. 10). Foam neoprene, which has largely replaced rubber for wet suits, has a thermal insulation of 1.48 clo in air, 0.76 clo in still water near the surface, and about 0.71 clo if the water is disturbed (Cooper, 1976). At greater depth compression of the gas cells reduces its thermal insulation. For excursions from an underwater chamber (see below) a dry suit may be electrically heated, but such a suit is bulky and difficult to put on (Bond, 1967).

Other SCUBA Equipment

Flippers are worn to increase efficiency and speed of movement through water. A weighted belt provides neutral buoyancy and can be jettisoned if rapid ascent is imperative. For any deep or prolonged dive a depth meter and watch are essential so that the limits of safety imposed by the tank capacity and by the hazards of pressure changes can be observed.

The Buddy System

Diving is dangerous and should never be performed alone. The buddy system (Edmonds, 1976), in which divers are paired off and each member of the pair must accompany his opposite number and be responsible for his well-being, is probably the most important safety precaution to observe.

Saturation Diving

Rapid ascent even from shallow depths is hazardous, and the duration of ascent, without exposing the diver to undue risk, is proportional to the depth and, up to certain limits, to the duration of the dive. For a very deep dive the time spent on ascent may be many times that spent on the bed of the ocean. For a deep dive it is more efficient to keep the diver at the required depth for a prolonged period, since once the tissues are fully saturated with the respiratory gases, any further time spent at this depth does not increase the time necessary for decompression (Vorosmarti, 1976).

For such "saturation diving," diving chambers on the principle of the traditional diving bell have been introduced, in which men have lived at a depth of 100 m for as long as 3 weeks, emerging with SCUBA equipment to explore and work in the neighborhood of the chamber (McInnis, 1966) (Fig. 11). Such a chamber carries its own gas supply and is in telephonic communication with a ship on the surface, from which it receives electricity for heating and lighting. Since the gas mixture in the chamber is usually at least 90 percent helium, the voice is distorted and communication may depend on other signals unless the speaker breathes air from a cylinder during short speech periods (Dickson and MacInnis, 1967).

Figure 11. Sealab II. (From J. B. MacInnis, *Scientific American, 214(3)*:24-33, 1966, by permission of Dr. MacInnis and the U.S. Navy.)

Armoured Suit

For work at very great depths without the risk of gas poisoning or the dangers of decompression on ascent the diver may be enclosed in an armoured suit at a pressure of 1 atm. Some limitation of movements is inevitable, but a suit developed in 1969 allows useful work to be done at a depth of 300 m for 4 hrs with a safety factor of 2, since it has withstood tests at 61 ata and has a gas supply sufficient for 8 hours of moderate activity (Bachrach, 1975).

Submarine

The modern nuclear submarine can remain submerged for more than 3 months at a time. Protection from ionizing radiation is adequate, the main danger being progressive accumulation of toxic gases (see below).

Bathyscaphe

For the deepest undersea exploration, about 3.2 km (2 miles) below the surface, the divers are enclosed in a very strong chamber (bathyscaphe), which can resist the intense external pressure of more than 300 ata.

COMPLICATIONS OF DEEP DIVING

Hypothermia

Loss of body heat is a limiting factor even in shallow-water diving and is greater, because of the lower water temperature, at greater depths. Heat loss is more rapid when nitrogen is replaced by helium in the gas mixture (see above). Suitable clothing (dry suit or wet suit) reduces heat loss and a long-term underwater shelter must be warmed. The treatment of hypothermia, when the diver has been brought to the surface, is immersion in hot water (Cooper, 1976).

Barotrauma

Sudden pressure changes may cause damage, particularly to the lungs and ears, if there is insufficient time for equalization of pressure with the environment.

Burst Lung

On ascent from 10 m (2 ata) the volume of the air in the lungs is doubled. A much smaller increase in volume, even from a depth of 1 m, if no air is expired, can cause permanent lung damage (Kidd and Elliott, 1975). Pulmonary alveoli are stretched and may rupture; so that air passes into pulmonary blood vessels, possibly causing cerebral air embolism, or into the pleural cavity, causing pneumothorax. The danger does not arise in breath-hold diving, where the air in the lungs merely expands to its original volume on ascent, but is present whenever air or another gas mixture has been supplied at more than atmospheric pressure.

Squeeze

Rapid descent compresses the air in the lung. At a depth of about 30 m (4 ata) this is compressed to the normal residual volume of the lung (Miles, 1966). The bony chest wall resists further compression and further pressure forces blood into the thorax, distending and even rupturing pulmonary blood vessels. Squeeze may occur, in an underwater fall, if the pressure of gas supplied is not increased sufficiently rapidly.

Ear and Sinus Barotrauma

The air-filled cavity of the middle ear communicates with the pharynx through the Eustachian tube, which is open only during the act of swallowing or during a forced expiratory movement with lips and nostrils closed (Valsalva maneuver). Unless the Eustachian tube is opened during descent to equalize the pressure in the inner ear with that in the environment and in the pharynx, the increasing external pressure forces the tympanic membrane inwards and may even rupture it. During ascent the gas in the middle ear expands as the external pressure falls, and the tympanic membrane may rupture outwards; but this is less common because positive pressure in the middle ear can usually open the Eustachian tube (Farmer and Thomas, 1975). Since middle ear and pharyngeal pressures can be equalized only through the Eustachian tube, any upper respiratory infection, which may temporarily block the tube, is a contraindication for diving.

Sinus barotrauma is second only to middle ear barotrauma as an occupational disease of divers (Fagan et al., 1976). It is due to blockage of the aperture through which an air sinus is normally in free communication with the nose, so that the pressure in the sinus is higher or lower than that in the rest of the body. Sinus barotrauma is commonest during descent, when the sinus blood vessels may become engorged with blood and rupture. The frontal sinus is most commonly affected. As a rule no treatment is necessary.

Decompression Sickness

The commonest complication of diving, decompression sickness, is due to too rapid ascent from a deep dive. The symptoms range from joint pains ("bends"), respiratory distress ("chokes"), and vertigo ("staggers") to paralysis and even death. Although any diver is at risk after a long deep dive, there is considerable individual variation in susceptibility. In general obese people are more susceptible than thin (Boycott and Damant, 1908) and older than younger (Hill, 1912).

Cause

The volume of a gas which dissolves in a liquid is directly proportional to the partial pressure of the gas in contact with the liquid and to the solubility of the gas in the liquid (Chapter 2). Consequently, at high pressures more of the respiratory gases are dissolved in the tissues than at sea level. Since nitrogen is particularly soluble in lipids the adipose tissue and nervous tissue, which are rich in lipids, take up more of this gas than other tissues. Equilibrium is not reached at once, the time to saturation of the slowest tissues being many hours (Boycott et al., 1908).

On ascent, as pressure falls, the gases dissolved in the tissues are given off into the blood and pass from the blood into the air which is expired. If the rate of evolution of gas from the tissues is too rapid, bubbles form, which may cause pains in the joints, breathlessness, or even cardiac arrest from air embolism. Oxygen seldom causes these disorders, because it is readily utilized by the tissues; nitrogen is more dangerous, especially because of its high solubility in lipids, which may lead to bubble formation in the nervous system, with laceration of nerves and consequent paralysis. This occurs most frequently in the thoracic region of the spinal cord (Strauss, 1976). The bubbles may also cause clumping of blood platelets and red corpuscles and extravasation of fluid, leading to circulatory failure (Elliott and Hallenbeck, 1975).

Diagnosis

The diagnosis of decompression sickness is usually obvious from the history and symptoms but, if an ultrasonic Doppler flow meter is available, may be confirmed by detecting bubbles in the superficial veins or in the heart (Spencer, 1976). Such bubbles may not be accompanied by symptoms but are nevertheless an indication for immediate recompression.

Prevention

Decompression sickness is avoided by slow decompression in stages after a deep dive. The general principle is to raise the diver to a depth at which the pressure is half that to which he was exposed at the foot of the dive, allow time for equilibration, and then repeat the process. The decompression tables drawn up on this principle in 1907 by Haldane (Boycott et al., 1908) are still the basis of present-day practice, with modification for the greater depth and duration of diving since Haldane's time (Bornmann, 1967; Workman and Bornmann, 1975). The reduction of pressure at each stage of decompression and the duration of the stage depend on the gas mixture, the pressure, and the duration of exposure to pressure. Desaturation is four times faster with helium than with nitrogen (Hempleman, 1975). Since decompression times (times to half saturation) range from 5 to 75 minutes for different tissues (Taylor, 1965), and it is not practicable to wait for equilibrium in the slowest tissues, a compromise may be adopted, at which not more than 5 percent of divers will experience symptoms on decompression (Spencer, 1976).

If the diver must be brought to the surface rapidly in an emergency, he is immediately recompressed in a compressed air chamber on the support ship, where gradual decompression is then carried out. A diver must be within reach of such a chamber for at least 12 hours after a deep dive, because the onset of symptoms of decompression sickness may be delayed; and he should not repeat such a dive within 24 hours since a variable amount of nitrogen remains in the tissues for at least as long as this. Modern practice in saturation diving is to bring the diver rapidly to the surface in a small pressure chamber, from which he is transferred to a larger compressed air chamber on the ship, so that the many hours of decompression may be passed in more comfort and safety than deep in the ocean (MacInnis, 1975).

Treatment

The treatment of decompression sickness is immediate recompression to 2.8 ata in pure oxygen, or if this is not available, to 6 ata in air followed by gradual decompression (Kidd and Elliott, 1975). Circulatory failure is treated with intravenous plasma or dextran (Kindwall, 1976b).

Gas Intoxication

Toxicity of the respiratory gases at high pressure limits the depth of diving on compressed air. For deep dives helium is usually substituted for nitrogen, and the proportion of oxygen in the gas mixture is reduced (see above); but even helium becomes toxic at very high pressure.

Nitrogen Poisoning

Nitrogen poisoning may occur breathing air at any depth below 30 m (4 ata) and is made worse by high pressure of oxygen or carbon dioxide and by exercise. The symptoms, described by Cousteau (1953) as *l'ivresse des grandes profondeurs* (rapture of the depths) are reduced self-control, slowing of mental processes, and overconfidence; they resemble intoxication with lysergic acid diethylamide (LSD) rather than with alcohol (Bennett, 1967). The diver may act foolishly, for instance tearing off his mask. Frequently repeated exposure increases resistance to nitrogen intoxication (Bennett, 1965).

To avoid nitrogen poisoning a mixture of 9 percent oxygen and helium is commonly used instead of air for deep dives.

Oxygen Poisoning

Pure oxygen even at 1 atm, causes irritation and inflammatory changes in the respiratory passages and lung in about 12 hours. At a pressure of 2 atm these symptoms appear in 3 to 6 hours (Wood, 1975). Oxygen at this pressure may also cause oxygen poisoning, the characteristic symptoms of which are giddiness and convulsions. Tingling sensations in the fingers and toes and tremor of the lips usually occur at an early stage of oxygen poisoning and are useful warning signs; if the warning is ignored convulsions may follow.

Individual susceptibility to oxygen poisoning varies widely; some individuals may suffer from it breathing air at a depth of 30 m (less than 1 atm of oxygen), but it is almost inevitable at 100 m (more than 2 atm of oxygen) (Miles, 1966). There is no treatment, except reducing the partial pressure of oxygen.

High Pressure Nervous Syndrome

During very deep dives on a helium-oxygen mixture the diver may suffer from the high pressure nervous syndrome (HPNS); he experiences tremor of the limbs and dizziness, which may progress to convulsions, coma and death. This is likely to occur at pressures in excess of 20 ata, especially if the descent is rapid (Bennett, 1976; Kindwall, 1976a). The addition of nitrogen (12-14%) to the gas mixture reduces the liability to HPNS without causing nitrogen intoxication (Bennett, 1975); on such a mixture, pressures in excess of 60 ata can be tolerated without disabling symptoms.

Aseptic Necrosis of Bone

A long-term danger of diving, even to shallow depths, is necrosis of bone. Radiological evidence of this is commonly found in professional divers (Ohta and Matsungana, 1974; Williams and Unsworth, 1976). Patchy areas of destruction are seen, especially in long bones and often near the joints, which may be involved and suffer permanent damage (Walder, 1976). The necrosis has been attributed to repeated blockage of small arteries in the bone by gas bubbles (Williams and Unsworth, 1976). There is no effective treatment.

SUBMARINES

The steel wall of a submarine gives protection against the external pressure of the ocean so that the air inside can be of normal composition and at a pressure of 1 atm. This involves constant monitoring of the air, removal of carbon dioxide, and replacement of oxygen. During prolonged submersion, purification of the air is necessary to prevent accumulation of contaminants; it is passed through burners to convert carbon monoxide and hydrocarbons to carbon dioxide, through scrubbers to remove carbon dioxide, and through charcoal filters to remove odors, dust, and bacteria. The oxygen level is kept at at least 18 percent, the carbon dioxide level not above 1.5 percent, and the level of carbon monoxide not above 50 parts per million (Miles, 1966).

Submarine Escape

Submarine escape involves rapid compression in the escape chamber and rapid ascent from there to the surface. The compression is so brief that there is no danger of decompression sickness, but the danger of barotrauma is considerable. Escape is started with the lungs full of air, which the subject exhales continuously, starting with a gentle flow and increasing it as the gas in his lungs expands (Miles, 1966). In some escape devices air from a life-jacket may be inhaled to allow an occasional brief inspiration during ascent (Weite et al., 1967).

DROWNING

An obvious danger of all underwater activity is drowning, "asphyxial death due to immersion in water" (Glaister and Rentoul, 1966). Although asphyxia occurs when either fresh or sea water is inhaled, the lethal process differs in the two cases. In fresh water osmotic pressure attracts a massive volume of water into the blood, leading to dilution of the blood, cardiac arrhythmia and sudden death. In the sea, water is attracted from the blood into the fluid in the air spaces in the lungs, and the blood becomes more concentrated, leading more slowly to cardiac failure. In some cases of drowning, spasm of the larynx prevents the entry of water into the lungs and death is from asphyxia, but the spasm is seldom sufficient to prevent some inhalation of water (Keatinge, 1969).

Treatment

The first aid treatment of drowning is immediate artificial ventilation, which may be performed by the mouth-to-nose or the mouth-to-mouth method, even before the victim is brought ashore (Sloan, 1979). The air passages must be kept clear of any obstruction, and artificial ventilation is continued, for hours if necessary, as long as the heart is still beating. Even a mild case should be under observation in a hospital for at least 24 hours after apparent recovery, since delayed pulmonary complications are not uncommon.

CHAPTER 6

MOUNTAINS

ALTHOUGH man can survive on the top of the highest mountain (8,848 m) as was demonstrated by the first successful ascent of Mount Everest (Hunt, 1953), he is subjected to increasing stress with increasing altitude, due to reduced atmospheric pressure and correspondingly reduced partial pressure of oxygen (Fig. 5). Susceptible individuals may suffer from mountain sickness (see below) at altitudes over 3,000 m; and the highest permanent residence, at Auconquilca in Chile, is 5,340 m above sea level (Fig. 12). Less than 1 percent of the world's population live above 3,000 m (Ward, 1975), the limiting factors being frozen soil and lack of oxygen. High altitude natives show some adaptation to their environment and acclimatization of visitors takes place for several weeks after the ascent.

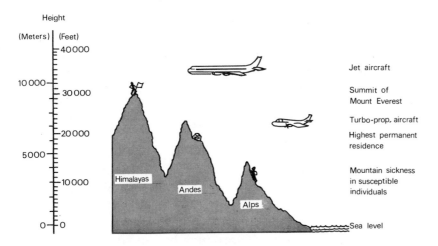

Figure 12. Terrestrial and supraterrestrial altitudes. (Reprinted with permission from A. W. Sloan: *Physiology for Students and Teachers of Physical Education*. London, Arnold, 1969.)

STRESSES ON HIGH MOUNTAINS

On high mountains man is subjected not only to oxygen lack but also to reduced humidity, intense cold, and strong ultra-violet irradiation.

Hypoxia

Severe oxygen lack causes loss of appetite and insomnia, leading to progressive physical and mental deterioration. The loss of appetite usually leads to undernutrition, and even if sufficient food is eaten, impaired metabolism may still result in loss of weight (Consolazio et al., 1972). Considerable weight loss is usual in members of high altitude climbing expeditions, even when an adequate and palatable diet is provided (Cerretelli, 1976).

Lack of oxygen limits the performance of physical work, which depends on the rate at which oxygen can be taken up by the lungs and carried to the active muscles. At altitudes above 1,500 m work capacity is reduced by about 3 percent for every additional 300 m in unacclimatized individuals and by about 2 percent in the acclimatized (Ward, 1975). At the Olympic Games in Mexico City (2,380 m) the performance of athletic events lasting about 4 minutes was reduced by 3 percent and of those lasting about an hour by 8 percent (Shephard, 1973). For some shorter events, not dependent on continuous intake of oxygen, performance was improved because of the diminished air resistance.

The less oxygen there is in the inspired air the greater is the work demanded of the respiratory muscles to oxygenate the blood. At about 9,000 m the maximum respiratory effort would be just sufficient to provide the oxygen required by the respiratory muscles; no other exertion is then possible without an additional supply of oxygen (Matthews, 1954).

Mental functions are affected by hypoxia. At altitudes above 3,600 m judgment and memory may be impaired, and there is slowing of reflex responses (McFarland, 1972).

Light sensitivity is impaired at altitudes above 1,200 m and may be reduced by 50 percent at 4,900 m. Dark adaptation is the visual function most affected by hypoxia (McFarland, 1972).

Hearing is unaffected except at very high altitude.

Reproduction also is affected. When the Spanish Conquistadores founded the city of Potosi at an altitude of 4,000 m in the Andes, with a population of 100,000 native Indians and 20,000 Spaniards, the Indians reproduced normally, but no Spanish child was reared for fifty-three years (Clegg, 1978). In visitors to high altitude there is a marked decrease in sperm count and sperm motility and an increase in abnormal forms (Donayre et al., 1968). In high altitude residents menarche is delayed and menstrual abnormalities are common (Heath and Williams, 1977). In communities resident above 3,600 m there is a high proportion of congenital abnormalities (McFarland, 1972).

There are relatively few old people in most high altitude communities, possibly because of progressive failure of adaptation to hypoxia with advancing age (Clegg et al., 1970).

Dehydration

Loss of water from the lungs is high on mountains because the inspired air is usually of low humidity and pulmonary ventilation is high due to hypoxia, to which may be added the response of the respiratory center to muscular exercise (Chapter 2). Sweating, due to the exercise, augments water loss and the thirst mechanism fails to promote adequate replacement (Pugh, 1964). Dehydration increases the viscosity of the blood, which increases the load on the heart.

Cold

Air temperature falls about 1°C for each 150 m of altitude (Heath and Williams, 1977). On high mountains, as in a polar environment (Chapter 4), high winds frequently add to the cooling power of the environment. Unless suitably clad, the mountaineer is liable to suffer from hypothermia or from local cold injury.

Ultraviolet Irradiation

The ultraviolet rays of sunlight, relatively unfiltered by the

atmosphere and reflected as much as 90 percent from snow or
ice (Heath and Williams, 1977), can cause sunburn of unprotected skin and retinal damage ("snow blindness") as in a polar
environment (Chapter 4).

ADAPTATION TO HIGH ALTITUDE

Individual Adaptation

Physiological Adaptation

Adaptation to high altitude involves a number of physiological changes: increase in pulmonary ventilation, changes in the
circulatory system, increase in the oxygen carrying capacity of
the blood, and increase in cellular utilization of oxygen. Work
capacity increases with acclimatization (Maher et al., 1974;
Sirotinin and Matsynin, 1977). Acclimatization usually nears
completion in 5 weeks but there is wide individual variation.
Completeness of acclimatization should be assessed not in relation to changes in the individual from the parameters at sea
level but in terms of similarity to the parameters of the native
population (Velazquez, 1970).

RESPIRATION. Hypoxia causes an immediate increase in pulmonary ventilation (Chapter 2). This flushes out carbon dioxide and lowers the carbon dioxide content of the blood,
making it more alkaline; but this change is checked by buffer
mechanisms in the blood and equilibrium is restored by increased elimination of bicarbonate in the urine. The carbon
dioxide tension in alveolar air is stabilized at a lower level, and
pulmonary ventilation is stabilized at a volume above the
normal for sea level (Chiodi, 1963). The immediate increase in
pulmonary ventilation on ascent to high altitude is presumably
due to hypoxic stimulation of carotid and aortic chemoreceptors, the sustained response to increased sensitivity of chemoreceptors to carbon dioxide (Kellogg, 1963).

CIRCULATION. The resting heart rate is increased during the
first week of acclimatization to altitude, after which it returns
to the previous level as other adaptive mechanisms are established (Luft, 1972). The systemic arterial blood pressure rises

slowly and remains at the higher level, due to increased viscosity of the blood (see below); and the pulmonary arterial pressure is high, due to this and also to the vasoconstriction of pulmonary blood vessels, which takes place in response to hypoxia (Comroe, 1974). There is an increase in the number of capillaries in the active tissues (Hurtado, 1963).

BLOOD. The oxygen-carrying capacity of the blood is increased by an increase in its hemoglobin content due to an increase in the number of red blood corpuscles, which takes place within two hours of reaching high altitude (Heath and Williams, 1977). At first the red cells are flushed out of reservoir areas by the more active circulation; but the high cell count is maintained by a hormone, erythropoietin, formed from a plasma globulin by the action of renal erythropoietic factor liberated by the hypoxic kidney, which stimulates the bone marrow to produce more blood corpuscles (Ganong, 1977). At moderate altitudes the same volume of oxygen may be carried in the blood as at sea level (Cerretelli, 1976); it is carried by a greater mass of hemoglobin though this is only partly saturated with oxygen (Fig. 13). There is also an increase in the concentration of diphosphoglycerate (DPG) in the erythrocytes, which promotes the liberation of oxygen from hemoglobin in the tissue capillaries (Luft, 1972). The increase in blood volume due to the increased number of red cells is partly compensated by a reduction in plasma volume due to dehydration and to a shift of water into the cells (Consolazio et al., 1972). The high concentration of cells in the blood increases its viscosity and thereby raises the arterial blood pressure and increases the load on the heart (see above). Blood values return to normal in about 40 days after return to sea level (Cerretelli, 1976).

TISSUES. The capacity of active tissues to utilize oxygen increases with acclimatization to high altitude. The number of mitochondria and the concentration of myoglobin increases in the cells (Hurtado, 1971), and there is also a greater capacity for anaerobic metabolism, though this is limited by reduced tolerance of the lactic acid produced due to the reduced blood alkali (Shephard, 1974).

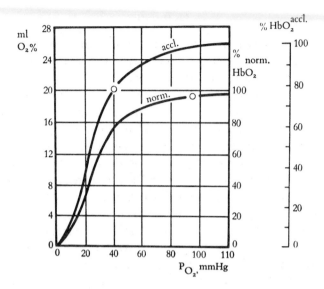

Figure 13. Oxygen dissociation curves of a sea-level resident (norm) and of an individual acclimatized to 5000 m altitude (accl). The open circles indicate the arterial oxygen content at atmospheric pressures of 760 mm Hg and 400 mm Hg respectively. (Reprinted with permission of P. Cerretelli and G. Monzino: *La Spedizione Italiana all' Everest*. Verona, Monzino, 1976.

Behavioral Adaptation

CLOTHING. The ideal clothing for mountaineering should provide adequate thermal insulation with the minimum interference with body movement. Less is worn during climbing, when heat production is high, than at rest. Like arctic clothing it should be windproof but permeable to water vapor so that sweat can evaporate. On snow-covered mountains sungoggles must be worn.

FOOD. A palatable high-energy diet is important in maintaining health when climbing high mountains. Mountaineers often express a preference for sweet foods, and a high carbohydrate diet may confer some protection against acute mountain sickness (see below). The daily fluid requirement on a high altitude climb is about 5 liters (Heath and Williams, 1977).

OXYGEN. The capacity for physical work on a high mountain is greatly increased by the provision of oxygen (Pugh, 1965). Most successful Everest expeditions have been dependent on an additional supply of oxygen carried by the climbers.

Racial Adaptation

The high altitude natives who have been most intensively studied are the Quechua Indians in the Andes and the Sherpas in Nepal. Both show some degree of adaptation to their environment. An interesting comparison has also been made of Amharas living at different altitudes in Ethiopia.

Quechua

The Quechua Indians have been living in the Andes, according to archaeological evidence, for thousands of years (Heath and Williams, 1977). They are lean and of small stature; and those who live at high altitudes have large chests, which may be regarded as a specific adaptive response to chronic hypoxia (Frisancho and Baker, 1970; Baker, 1978). Their resting respiratory rate is fast, resting heart rate slow, and they have a high red cell count in the blood (Hock, 1970). Those living at higher altitude have increased cardiac efficiency (Moret, 1977), and their exercise tolerance is higher and lactate production lower than in lower altitude natives (Hurtado, 1971). Their pulmonary blood pressure is normally high.

Some Quechua live at the upper limit of altitude for adaptation. The mine workers at the sulphur mine at Auconquilca in Chile, at an altitude of 5,760 m, prefer to live in a settlement at 5,340 m, though this involves a climb of 420 m each day to go to work (Keys, 1936). Living in a camp at a higher level they complained of insomnia and loss of appetite.

The exercise tolerance of the Quechua is no better than that of trained and acclimatized Caucasians, which indicates that genetic factors are relatively unimportant (Mazess, 1969).

Sherpas

The Sherpas migrated from Tibet into Nepal about 200 years ago (Baker et al., 1977) and live in scattered villages at an average altitude of 3,800 m (Fürer-Haimendorf, 1964). They are a short people with large chests, well-developed calf-muscles, and very little body fat (Sloan and Masali, 1978); all of these features may be considered adaptations to a life of carrying heavy loads in high mountains, there being no wheeled transport and little use of pack animals in Sherpa villages. The growth period of young Sherpas is long, and there is no obvious adolescent spurt (Baker et al., 1977). Their respiratory response to carbon dioxide is blunted, as in other high altitude natives, but this phenomenon and the large lung capacity are acquired, not inherited, characteristics; these characteristics are not found in descendants of high-altitude natives born and reared at sea level (Lahiri, 1971). The resting metabolism of the Sherpas is similar to that of Nepalese plainsmen, but the Sherpas have a higher work capacity at high altitude (Das and Saha, 1967; Kennter, 1969; Buskirk, 1978).

Amharas

In Ethiopia three genetically similar populations, mostly of Amharas, living at different altitudes, have been compared (Harrison et al., 1969). Those at 3,000 m and at 3,700 m are heavier, though not taller, and have larger chests than those at 1,700 m. Respiratory capacity is greater in the higher altitude groups, but the red cell count is not significantly raised. Body fat is low in all groups.

FAILURE OF ADAPTATION TO HIGH ALTITUDE

Acute Mountain Sickness

Mountain sickness was experienced by the Spanish Conquistadores in Peru and described by the Jesuit, Acosta, in 1590 (Bert, 1878). Acute mountain sickness (*"soroche"*) is a complication of rapid ascent to 3,000 m or higher, allowing insufficient time for adaptation to take place. Since speed of ascent is a critical factor, it is commonest in those who reach high altitude by wheeled transport or by aeroplane. On the 1971 International Himalayan Expedition the only cases of acute mountain sickness were in two visitors to the base camp (Steele, 1971). The first symptoms may be euphoria and irresponsibility, but further symptoms usually appear within the first 24 hours. These include breathlessness, palpitation, giddiness, headache, insomnia, weakness, and loss of appetite (Heath and Williams, 1977). The more severe cases suffer from vomiting and oedema, especially of the face and of the feet and legs. The output of urine is reduced.

Acute mountain sickness is probably due to cerebral oedema caused by the shift of body water into the cells, which results from hypoxia (see above), and to sodium and water retention due to hypersecretion of adrenal corticosteroids and vasopressin (Singh et al., 1969). Hypoxia also increases cerebral blood flow and permeability of cerebral capillaries.

The best mode of prevention of acute mountain sickness is gradual ascent with staging camps to permit acclimatization at successively higher altitudes. A high carbohydrate diet is believed to confer some protection (Ward, 1975), and a drug which increases the output of urine, e.g. acetazolamide, may be given in advance to those known to be susceptible (Cerretelli, 1976).

The symptoms of acute mountain sickness usually pass off in about three days; if they persist the patient should be brought down to lower altitude. The treatment is rest, sedation, and administration of a drug to increase the output of urine. The drug of choice in this case is frusemide, and severe cases should receive also morphine and betamethasone, which potentiate the action of frusemide (Singh et al., 1969). A small dose of morphine also has a useful tranquilizing effect without undue depression of respiration. Oxygen therapy is disappointing and may delay acclimatization (Heath and Williams, 1977).

High Altitude Pulmonary Oedema

The most dramatic and dangerous complication of mountaineering is acute pulmonary oedema, manifested by blueness of the skin and mucous membranes; intense breathlessness; and cough with frothy, blood-stained sputum (Cerretelli, 1976). It is commonest at moderate altitudes (2,600-3,700 m) (Shephard, 1973). Rapid ascent and strenuous exercise are precipitating factors and the condition is common in high altitude natives returning from a spell at lower altitude (Hurtado, 1971).

High altitude pulmonary oedema is always associated with a high red cell count and high pulmonary blood pressure, hence its common occurrence in individuals previously acclimatized to high altitude. The explanation is obscure because the pulmonary hypertension is due to vasoconstriction of pulmonary arterioles, and there is no evidence of raised pulmonary capillary or venous pressure. Severinghaus (1971) attributes the oedema to leakage of fluid through the distended walls of the branches of the pulmonary arteries; Heath and Williams (1977) believe it is due to blockage of the capillaries. However it may be caused, the pulmonary oedema presents an obstacle to the passage of oxygen from alveoli to pulmonary capillaries, resulting in respiratory distress and blueness of the skin and mucous membranes.

The treatment, which must be started without delay, is to administer oxygen and carry the patient down to lower altitude, preferably to sea level. Frusemide is administered, and it may be expedient (though paradoxical) to couple this treatment with administration of fluid to prevent shock and eliminate surplus bicarbonate (Cerretelli, 1976).

Chronic Mountain Sickness

Chronic mountain sickness (Monge's disease) is failure of adaptation after acclimatization to high altitude (Monge, 1943). It is commonest in young men (Penaloza et al., 1971). The patient is blue and his exercise tolerance is greatly reduced. The red cell count is very high and pulmonary hypertension is severe. The cause seems to be reduced sensitivity to carbon dioxide, leading to inadequate ventilation of the lungs, and consequent hypoxia (Hurtado, 1964; Pugh, 1965). Recovery is usually rapid on descent to sea level, but the susceptible individual is liable to a recurrence of the condition on returning to high altitude.

Cold Injury

Survival on a mountain, even at relatively low altitude, usually depends mainly on maintenance of body temperature. The greater the altitude the greater the danger of cold injury from low temperature and the cooling effect of high winds. Fatigue is an additional complication, since it may be impossible for the tired climber to maintain enough physical activity to counter the loss of heat.

Cold injury may be general (hypothermia) or local (frostbite) (Chapter 4). Preventive measures are wearing suitable clothing, eating enough energy-giving food, avoiding physical exhaustion, and making use of any available shelter. The best restorative fluid for a conscious patient is hot soup (Taylor, 1972). The body, including the head, should be covered with thermally insulating material and the patient should be carried down the mountain. Frostbite should be treated as soon as possible with general (not local) warming, antibiotics, and possibly anticoagulants (Cerretelli, 1976).

High Altitude Deterioration

The loss of weight and of energy, which is almost inevitable above 5,500 m, is due to depression of appetite and of thirst (Ward, 1954). Since it is likely to occur even if efforts are made to promote the intake of food and of water, attempts at acclimatization above this altitude are ineffective and the final assault on a mountain above 5,500 m high should be made with as little delay as possible. Supplementary oxygen slows the rate of deterioration (Bourdillon, 1954).

CHAPTER 7

AVIATION

THE stresses encountered in aviation include those of mountaineering (reduced atmospheric pressure, hypoxia, and cold), which are usually more extreme because aircraft can reach higher altitudes and reach them more rapidly than a climber. There is insufficient time for adaptation to occur. Propellor-driven aircraft can reach the height of Mount Everest in less than 10 minutes and jet aircraft normally fly above this level. Additional stresses are imposed by sudden changes of speed or direction.

Another hazard is introduced by the high speed of modern aircraft. By the time an object is seen directly in front of the aircraft, it may be too late to avoid it. At supersonic speeds collision may occur even before the object is recognized (Whiteside, 1965c).

The high speeds of jet aircraft also result in disturbance of human circadian rhythms after long flights across time zones. After a flight of 5,100 km between New York and London the physiological rhythms are out of phase with the day-night sequence by about 6 hours, and readjustment takes several days (Strughold, 1971b).

STRESSES IN AVIATION

Reduced Atmospheric Pressure

Rapid ascent in an unpressurized aircraft, like rapid ascent from deep water (Chapter 5), causes rapid expansion of gases inside the body as the pressure falls. The dangers are less in aviation than in diving, because the pressure change with altitude is much more gradual; only at an altitude of 5,500 m is the atmospheric pressure reduced to half what it is at sea level (Fig. 5). As in ascent from deep water, when the external pressure falls to half its previous level the gas in the body expands to twice its original volume, but in aviation there is usually plenty of time for the lung gas to be exhaled and for gas in the middle ear and air sinuses to pass out, unless there is any blocking of the orifices. Expansion of gas in the alimentary canal is not sufficient to cause local damage but may cause abdominal discomfort and even hinder respiration by upward pressure on the diaphragm.

Hypoxia

As the atmospheric pressure diminishes with altitude so does the partial pressure of oxygen (Fig. 5). At about 11,700 m, where the atmospheric pressure is about one fifth that at sea level, the partial pressure of oxygen in the atmosphere is one fifth that at sea level, but its partial pressure in the lungs is lower because the partial pressures of water vapor and of carbon dioxide in the alveoli of the lungs are the same as at sea level (Strughold, 1971a). Even pure oxygen, supplied at 0.2 atm pressure, will oxygenate the blood less than air at 1 atm. At higher altitudes oxygen must be supplied under positive pressure to prevent hypoxia. Since aviation does not allow sufficient time for adaptation to high altitude, additional oxygen should be provided for any flight above 3,000 m (Ernsting and Stewart, 1965).

Cold

Ambient temperature falls with increase in altitude (Chapter 6) up to an altitude of about 10 km. Between 10 and 30 km the mean temperature is about $-55°C$ (Luft, 1974).

Acceleration

Unless he looks at some stationary object, man is not conscious of movement at constant velocity, but he is aware of changes in velocity (acceleration) whether linear or angular. Sudden changes in velocity are a complication of aviation.

The unit of acceleration adopted in aviation is that of gravity (g). The force of gravity makes objects fall towards the earth (unless impeded by wind resistance) at an acceleration of 981 cm/sec^2. An aviator flying horizontally is subjected to this downward force. A change of direction imposes an acceleration towards the center of the turn, which is measured in g units and exerts a corresponding force in the opposite direction. An aviator pulling out of a dive or executing an inward turn undergoes headward acceleration and is subjected to force in the direction from head to feet (Fig. 14). This is referred to as +g. When the aircraft turns down suddenly or executes an outward turn the force is towards the head (-g). Acceleration has physiological effects on the body, depending on whether it is positive, negative, or transverse (see below).

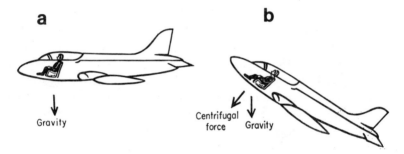

Figure 14. Forces exerted on aviator: a, flying level, and b, pulling out of dive. (Reprinted with permission from A. W. Sloan: *Physiology for Students and Teachers of Physical Education.* London, Arnold, 1969.)

Noise

The noise from a jet engine at a distance of 15 m may be 140 dB* above the threshold of hearing, and engine noise inside a propellor-driven aircraft can exceed 100 dB (Guignard, 1965). Even in a sound-proofed cabin it may exceed 60 dB.

---

*If the intensity of one sound is ten times that of another it is 1 Bel louder; if it is 100 times (10^2) that of the other it is 2 Bels louder. Since the Bel is an inconveniently large unit the decibel (dB) is preferred 10 dB = 1 Bel). Sound intensity is commonly expressed in dB relative to the international standard threshold of audibility (0.0002 $dynes/cm^2$ at 1000 Hz).

PROTECTION OF THE AVIATOR

Individual Protection

Protection of the aviator during high altitude flight involves provision of oxygen and of suitable clothing. For aerial combat some protection against the effects of positive acceleration is also desirable. Sun glasses or a visor are other important items of the pilot's equipment.

Oxygen Supply

Oxygen may be supplied to face-masks from cylinders of compressed oxygen. The modern practice for high altitude flight is to pump air into the cabin (see below), but a reserve of oxygen should be available to individual face-masks for flying crew and, in the event of failure of the air pressure system or serious cabin leakage, to passengers. In military aircraft individual oxygen supply is essential because of the greater danger of cabin damage and of having to escape from a damaged aircraft at high altitude; the supply must be adequate to maintain normal oxygenation of the blood up to an altitude of 10,000 m, with a peak flow rate of 250 liters/min (Ernsting, 1965). An open circuit system is more satisfactory than a closed circuit, and a demand system is more economical of gas than continuous flow. The mask and its harness should be suitable for pressure breathing if necessary.

Clothing

Cold was a major complication of flying in an open cockpit, and the protective clothing worn by early aviators was sufficiently bulky to interfere with movement. The practice today is to warm the air passing into the cabin of the aircraft, so no special clothing is necessary in civil aircraft; but combat clothing must have sufficient thermal insulation to protect the aviator against the extreme cold encountered on damage to the cabin or on abandoning the aircraft. For normal wear the flying suit may be air-ventilated or incorporate a liquid cooling system.

Protection from Acceleration

The main protection necessary is against high positive acceleration in military aircraft, which drains the blood into the lower part of the body, causing temporary blindness and loss of consciousness (see below). The prone or supine position would convert the acceleration of a tight turn into a transverse vector and avoid blackout, but it is difficult to control the aircraft from this position. In the seated posture crouching or raising the legs gives some protection; a raised rudder bar may be provided, but a more effective procedure is to wear an "anti-g suit" (Howard, 1965b). This has bladders over abdomen, thighs, and calves, which are automatically inflated with air when the aviator is exposed to positive acceleration; it raises the threshold for black-out about 2 g. The immediate action of the anti-g suit is to increase the resistance to blood flow in the lower part of the body; during prolonged acceleration it reduces pooling of blood in the leg veins and transudation of fluid from the blood into the tissues of the lower limbs.

Protection from Noise

The best protection against noise is to make less noise. Aerodynamic noise may be reduced by structural modification of the aircraft and engine noise by efficient mufflers. Internal noise is reduced by efficient sound-proofing of the cabin and personal protection is by ear-muffs. For speech communication in intense noise microphones and earphones, linked with an appropriate amplifier, are designed to reduce the masking effect of background noise (Kryter, 1970).

Pressure Cabin

For all high altitude flight the present day practice is to maintain the atmosphere in the cabin at a comfortable level of pressure, temperature, and humidity. Air is pumped into the cabin, and since humidity at high altitude is very low and dry air is irritating to the respiratory epithelium, water is added to create 30 to 50 percent relative humidity. The air may be warmed by heat from the engine or, if necessary, cooled to counteract the heating effect of compression. At very high speeds cooling is necessary because of heating of the aircraft fuselage by friction with the atmosphere (Preston, 1975).

Adequate ventilation demands some recirculation of cabin air. An air velocity of at least 4.5 m/min is required for comfort (Brown, 1965b), and offensive odors must be reduced by efficient filters, especially on long flights.

It is not practicable to maintain sea level pressure at high altitude because of the high pressure differential on the cabin wall and the danger of explosive decompression (see below). Passenger aircraft are usually pressurized to an equivalent altitude of 2,500 m when flying at 12,000 m. Military aircraft have a "cruise setting," similar to that of civilian aircraft, and a higher "combat setting," up to about 7,500 m, because of the greater danger of explosive decompression from damage to the cabin and the risk of having to abandon the aircraft (Brown, 1965b).

COMPLICATIONS OF AVIATION

Failure of any of the protective measures described above may result in disorders of function. These may be due to pressure changes, hypoxia, cold (Chapter 4), movement, and noise.

Disorders Due to Pressure Changes

Barotrauma

Barotrauma to ears, air sinuses, or lungs rarely occurs in aviation because the pressure changes are much more gradual than in diving (see above). Discomfort in the ears, which may be experienced during ascent or descent, is relieved by swallowing or by the Valsalva maneuver to open the Eustachian tube (Chapter 5). Blockage of the tube, for instance by an upper respiratory infection, greatly increases the danger of damage to the tympanic membrane.

Decompression Sickness

The commonest symptom of decompression sickness in aviation is "bends" (Chapter 5). Other symptoms include itching, "chokes," and "staggers" (Gemmill, 1943). Paralysis is rare, but circulatory collapse can occur and may be fatal (Fryer and Roxburgh, 1965). The treatment is rapid descent to sea level and, in severe cases, recompression to more than 1 atm in a pressure chamber. In aviation, as in diving, susceptibility to decompression sickness increases with age and with obesity.

The risk of decompression sickness can be reduced by eliminating some nitrogen from the body before the ascent. At sea level there is approximately 1 liter of uncombined nitrogen in the body of an adult man. If pure oxygen is breathed for half an hour before the flight about half of this nitrogen is eliminated (Armstrong, 1943).

Explosive Decompression

Very sudden decompression takes place if a large hole is blown in a pressurized cabin while the aircraft is at high altitude. Although this is a very rare occurrence in civil aviation it is an ever-present hazard in aerial combat. Explosive decompression causes a blast wave through the body and very rapid expansion of gas in the intestinal tract, respiratory passages and lungs, and middle ear. Sudden lung expansion is limited by inertia of the chest wall, but breathholding, or obstruction to expiration by a face-mask, increases the danger of rupture of the lungs or of air embolism. Fortunately all these complications are rare; the greatest danger is of ejection from the aircraft by the sudden outrush of air (Fryer, 1965).

Disorders Due to Hypoxia

Acute Hypoxia

The symptoms of hypoxia are the same in aviation as in mountaineering, (Chapter 6) but in aviation the onset is usually more rapid. The early symptoms, which may include euphoria and lack of judgment, may be missed and consciousness may be lost without warning. Death may occur from respiratory failure (Armstrong, 1943), but recovery is rapid if oxygen is provided promptly.

Night Blindness

The retinal receptors responsible for vision in dim light (the rods) are particularly susceptible to hypoxia, even of moderate degree, so night blindness (more commonly due to deficiency of vitamin A) is a complication of night flying unless additional oxygen is inhaled. For night flights above 1,200 m the pilot should have additional oxygen (Ernsting and Stewart, 1965; Whiteside, 1965b).

Disorders Due to Movement

Positive Acceleration

An aircraft pulling out of a dive or in a high speed turn may be subjected to a centrifugal force corresponding to an acceleration of as much as +5 g. At this acceleration the body weighs 5 times its weight at rest, the cheeks sag, the abdominal muscles sag, and the aviator cannot stand up or lift an object from the floor (Armstrong, 1943); if the acceleration is prolonged he becomes blind ("black-out") and loses consciousness. The centrifugal force pulls the blood into the lower part of the body, lowering the arterial pressure in the head and neck. There is narrowing of the visual field due to failure of retinal circulation and, when the pressure in the retinal artery falls below intraocular pressure (about 20 mm Hg), retinal circulation stops and the individual becomes blind (Howard, 1965a). If the acceleration is maintained, failure of the cerebral circulation causes loss of consciousness. Recovery after a short exposure is rapid and complete.

An average threshold for blackout is about +5 g for 5 seconds, but there is wide individual variation in tolerance of acceleration (Christy, 1971). The principal air forces have human centrifuges in which the threshold for blackout of potential pilots can be tested.

A lower acceleration can be tolerated for a longer time because some compensatory mechanisms come into play; these are constriction of blood vessels in the lower limbs and an increased heart rate, which raise the blood pressure in the head and neck, though not to the original level (Howard, 1965b). Owing to the inertia of the blood, a very high acceleration can be tolerated for a very short time, as in a crash landing.

Factors which increase susceptibility to positive acceleration include heat, hypoxia, and hyperventilation, of which hyperventilation is the most potent (Howard, 1965b). It follows that excitement or anxiety, which promote hyperventilation, increases the risk of losing consciousness. No acclimatization to acceleration takes place with repeated exposures, but tolerance can be increased by appropriate posture or by an "anti-g suit" (see above).

Negative Acceleration

Negative acceleration is rarer than positive, and human tolerance to it is low. Centrifugal force drives blood upwards into the head, causing flushing of the face, and the individual may become temporarily blind from extreme congestion of the retinal blood vessels. Loss of vision may be preceded by reddening of the visual field ("red-out"), due to retinal congestion or hemorrhage from the retinal vessels or to the centrifugal force pulling the lower eyelid up over the eye (Howard, 1965c). It is followed by loss of consciousness, due to the rise in cerebral venous pressure reducing cerebral blood flow, and possibly to reduced oxygenation of arterial blood from mechanical interference with breathing. An average threshold for loss of consciousness is about -4 g for 5 seconds (Howard, 1965c). Cerebral hemorrhage rarely occurs because the cerebral blood vessels are supported by a corresponding rise in the pressure of cerebrospinal fluid.

Transverse Acceleration

High transverse acceleration is seldom experienced in aviation except at a crash landing. Tolerance to transverse acceleration is as high as 12 g for 2-3 minutes (Balke, 1963); the limiting factor to positive transverse acceleration (force acting from front to back of the body) is difficulty in breathing due to compression of the lungs by flattening of the chest and by increased abdominal pressure raising the diaphragm (Howard, 1965d).

Irregular Movement

Most people are susceptible in some degree to motion sickness; the complaint is less common in aircraft than in ships, because the movements of aircraft are less regular and less sustained. Airsickness is usually caused by sudden vertical movements and may be augmented by fear of being sick (Whiteside, 1965a).

Since head movement is the main factor in motion sickness it should be avoided as much as possible. Drugs which may give relief include hyoscine hydrobromide, dimenhydrinate (Dramamine®), and cyclizine hydrochloride (Marzine®).

Disorders Due to Noise

Failure of Communication

Speech communication is interfered with by noise in the frequency range between 600 and 4800 Hz* (Parker and von Gierke, 1974); at 65 dB above threshold, speech communication is impaired even with voices raised (Kryter, 1970).

Deafness

Sound intensity of 135 dB above threshold causes pain and vertigo, and deafness is likely to result from repeated exposure to sound intensities above 85 dB (Guignard, 1965).

---

*Hz = Hertz (cycles/sec)

ESCAPE FROM AIRCRAFT

Ejection Seat

Leaving an aircraft flying at more than 290 km/hr is difficult and dangerous unless the aviator is forcibly ejected (Nuttall, 1971). Downward ejection is undesirable because of the high negative acceleration, and at low levels, there would be insufficient time for the parachute to slow the rate of fall before landing. Upward ejection must be sufficiently rapid for the aviator to be projected clear of the tail of the aircraft. The duration of ejection is too short for black-out to occur; the main danger is compression fracture of vertebrae from the sudden jolt (Kaplan, 1974), but the heart also is liable to damage (Krefft, 1974). At speeds above 185 km/hr the aerodynamic force encountered on leaving the aircraft is liable to cause injury unless the aviator is protected (Jones and Jones, 1965).

Parachute Descent

To reduce the time of exposure to cold and hypoxia free fall to about 3,000 m is desirable; this also reduces the opening shock of the parachute which, at altitudes above 7,500 m, is likely to damage both parachute and man (Glaister, 1965). Forward acceleration is rapidly reduced by wind resistance, and downward acceleration (initially 981 cm/sec^2) is reduced by wind resistance at lower altitudes to a terminal velocity of about 51 m/sec (120 mph) without a parachute and about 6.7 m/sec (16 mph) with the parachute open (Jones and Jones, 1965). A barometric mechanism opens the main parachute below 3,600 m (Glaister, 1965). The ejection seat is jettisoned either shortly after ejection or when the main parachute opens.

Even falling freely it takes more than 2 minutes to descend from 12,000 m to 3,000 m (Jones and Jones, 1965), during which time consciousness may be lost unless supplementary oxygen is inhaled. An emergency oxygen supply must therefore be part of the ejection seat-parachute complex.

Survival on Landing

An aviator who abandons his aircraft and achieves a successful descent to land in a hostile environment must be protected against this environment, whether it be land or sea,
tropics or arctic. Provision must also be made for location and
rescue and for water, food, and ancillary aids.

The problems of survival in extreme terrestrial environments
have been described in previous chapters. Flying clothing provides some thermal insulation, and for flights over cold seas
some form of immersion suit is an important item of clothing.
Location and rescue are aided by a personal HF radio trans-
mitter, signals from which can be picked up by the rescue
aircraft. Emergency rations are based on an energy requirement
of at least 2,000 kcal and at least 500 ml water per day (Whittingham, 1965).

CHAPTER 8

SPACE FLIGHT

AN extreme environment, which man has entered only in the past twenty years, is space outside Earth's atmosphere. Exploration of space is a sequel to the development of rocket technology during World War II, which made it possible to put artificial satellites into orbit around Earth and even to accelerate them to the velocity of more than 39,000 kph (24,000 mph), which is required to escape from Earth's gravitational field and launch out into space (Bullard, 1972). The first man to circle Earth in orbit was the Russian, Yuri Gagarin, in 1961. In 1962 John Glenn, in an American satellite, orbited Earth 3 1/2 times in 4 1/2 hours. In 1965 both Russian and American astronauts moved about in space outside their spacecraft. In 1969 the American astronauts, Neil Armstrong and Edwin Aldrin, were the first men to walk on the moon: the whole Apollo 11 mission, including about 15 hours spent on the moon, lasted about 8 days. In 1973-74 three American astronauts orbited Earth for 84 days in Skylab 4, and since then still longer flight has been achieved.

PHYSICS OF ATMOSPHERE AND SPACE

Exploration of the upper atmosphere dates from the invention of the hydrogen balloon in 1783. Early ascents reached altitudes of only a few thousand meters; but Auguste Piccard in 1913 entered the stratosphere; and in 1961 an American balloon with a sealed cabin reached an altitude of 34.6 km (21.5 miles). In 1963 an American manned aircraft flew above 100 km (Shelton, 1972).

The major regions of the atmosphere are the troposphere (up to 10 km), which can maintain life, the stratosphere (10-80 km), and the ionosphere (80-700 km), which merges into interplanetary space (Fig. 15).

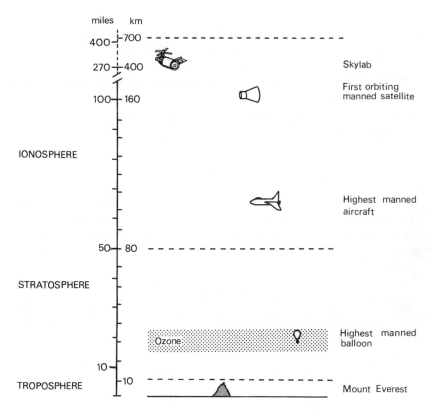

Figure 15. Manned exploration of the earth's atmosphere.

Composition

In the troposphere the atmospheric pressure falls progressively with increase in altitude (Fig. 5). In the troposphere the proportions of the gases are the same as at sea level, except for lack of water vapor at high altitude; but in the stratosphere hydrogen and helium form a significant proportion of the scanty gas molecules present. A layer of ozone, with its maximum density at about 26 km (Spells, 1965) is an important shield against solar ultraviolet radiation. In the ionosphere there is no effective atmosphere offering wind resistance to the passage of solid objects, but ionized particles reflect some radio waves and have a "greenhouse" effect, reflecting heat waves back to Earth.

Interplanetary space is empty except for scattered gas molecules (mostly hydrogen) and widely scattered particles of solid matter ranging in size from dust to large meteorites. Beyond 1,000 km above the geomagnetic equator Earth is surrounded by a zone of charged particles, the highest concentrations of which (the Van Allen belts) are at altitudes of 2,400 and 24,000 km (Shelton, 1972).

Temperature

In the troposphere atmospheric temperature diminishes with altitude (Chapter 7). In the stratosphere it is about −50°C over the poles and about -80°C over the equator, rising to 0°C at an altitude of about 60 km and falling to −80°C at about 80 km (Spells, 1965). In outer space there is no atmospheric temperature, because there is no atmosphere. The radiant heat of the sun in the region of Earth's orbit is about 2 cal/sec/cm², but the temperature of interplanetary dust is probably about −170°C (Hardy, 1964a).

STRESSES IN SPACE FLIGHT

The stresses encountered in space flight are those imposed by the inhospitable environment, by high acceleration and weightlessness, and by prolonged isolation.

Lack of Atmosphere

Earth's atmosphere protects it from meteorites, most of which are burnt up by friction before reaching the surface. The atmosphere also provides some shielding from ultraviolet and cosmic radiation. Serious damage to a spacecraft by a meteorite, even on a prolonged flight, is unlikely, but exposure to ionizing radiation is a hazard. The unfiltered ultraviolet rays of the solar spectrum would damage the unprotected eye.

The very low atmospheric pressure above the troposphere carries another hazard. The lower the pressure the lower the temperature at which water boils. At about 19 km altitude, where the pressure is less than 0.1 atm, the boiling point of water is 37°C, which is the core temperature of the human body (Chapter 1). Exposed to lower pressures than this the body fluids would boil (Luft, 1974).

Extremes of Temperature

Extremes of temperature are encountered in space flight, from the extreme cold of outer space to the intense heat generated by friction with the atmosphere during lift-off and reentry.

Ionizing Radiations

The ionizing radiations encountered in space include X-rays, Bremsstrahlung, gamma rays, and corpuscular radiations, all of which damage living cells. Stellar cosmic radiation fluctuates in an eleven-year-cycle, coincidentally with the frequency of sunspots; the next sunspot maximum is predicted for 1981. Solar radiation storms may, however, occur at any time.

Acceleration

The acceleration necessary to attain orbital flight is at least 8 g for about 2 minutes, which is near the limit of human tolerance even when applied transversely (Chapter 7). Deceleration on reentry, though controlled by retro-firing rockets, and finally by parachutes, is as high as or higher than the acceleration of take-off (Bullard, 1972).

Weightlessness

Weightlessness is experienced in orbital flight, when the angular acceleration is 1 g, and during the greater part of interplanetary travel. Human tolerance of this unusual state is better than was anticipated (see below).

Complications of weightlessness include the free movement of the body inside the spacecraft, unless some restraint is exercised; and the free movement of particles of solid matter, which may be inhaled. Absence of gravity also abolishes convection currents.

Isolation

The psychological effects of space travel so far have not proved serious. Only the earliest flights were undertaken by a single individual, and radio communication with Earth is maintained throughout the journey. Nevertheless, individuals in the very prolonged isolation of interplanetary travel could possibly develop psychological problems.

Inactivity

The confined space in a spacecraft limits the physical activity possible for its occupants. Prolonged inactivity causes muscle wasting and decalcification of the skeleton (Dietrick et al., 1948; Hawkins and Zieglschmid, 1975).

PROTECTION OF THE ASTRONAUT

Space Capsule

Since the environment of the upper atmosphere and of interplanetary space is quite incompatible with life, a manned spacecraft must incorporate a sealed capsule in which conditions compatible with life are maintained.

Atmosphere

The atmosphere in the space capsule should have a total pressure in the range 187-760 mm Hg: partial pressure of oxygen about 160 mm Hg; partial pressure of carbon dioxide below 8 mm Hg; relative humidity 40-60 percent; and no toxic contaminants (Karstens and Welch, 1971). Oxygen is carried in the liquid state and carbon dioxide is absorbed by lithium hydroxide (Brady et al., 1975). On prolonged flights the carbon monoxide released from metabolism of hemoglobin (about 10 ml/day) must be disposed of (Luft, 1974). Early American spacecraft had an atmosphere of pure oxygen at about 260 mm

Hg, whereas Russian spacecraft had normal air at 760 mm Hg (1 atm). Since pure oxygen, even at low pressure, is a fire hazard and may cause inflammation of respiratory epithelium and even patchy collapse of the lung, nitrogen is now added to the atmosphere in American spacecraft. In the later Apollo spacecraft the average gas mixture at lift-off was 65 percent oxygen, 29 percent nitrogen, 4 percent water vapor, and 2 percent carbon dioxide at a total pressure of 260 mm Hg (Sawin et al., 1977), but, due to gas leakage from the capsule and replacement with pure oxygen, the proportion of nitrogen diminishes progressively; after several days the atmosphere is almost 100 percent oxygen (Michel et al., 1975). A low pressure, though it does not require such a strong wall to resist the pressure gradient in outer space, introduces the risk of decompression sickness, unless pure oxygen is breathed prior to the ascent (Chapter 7).

In the absence of natural convection, air movement of at least 5 m/min should be provided artificially (Brown, 1965).

Temperature Regulation

The space capsule has effective thermal insulation and the heat produced by the astronauts and by the electrical equipment inside the capsule is dissipated during flight through radiators in the wall of the capsule, on the same principle as water-cooling of an internal combustion engine. For protection against the extreme heat of reentry the capsule reenters the atmosphere backwards; and the back of the capsule has a heat-protective shield, much of which is burnt off during reentry.

Protection from Ionizing Radiation

Some shielding from ionizing radiation (about 1 gm/cm^2 aluminium equivalent) is provided by the wall of the space capsule (Clark, 1964) and some additional protection by the space suit. Peaks of solar activity can be predicted and space travel should be avoided during these periods. Furthermore the orbit of a manned satellite should be planned to avoid as far as possible the Van Allen belts. During a burst of increased solar activity the astronaut should be inside the spacecraft and wearing a space suit. So far, on flights as long as 84 days, exposure to ionizing radiation has been well within the limits of safety (Bailey et al., 1977).

Protection from Acceleration

The very high acceleration necessary to get into orbit or to leave Earth's gravitational field can be tolerated only if it is applied transversely to the body. At take-off the astronaut is on his back on an individually moulded couch, with hips and knees flexed (Fig. 16). In this position he is seated upright when the spacecraft is in orbit.

Reentry imposes very rapid deceleration. The spacecraft is maneuvered to descend backwards so that the astronaut is forced into his couch instead of being flung out of it.

Adaptations for Weightlessness

Restraining straps are necessary to prevent the astronaut floating about in the cabin. For sleeping the astronaut may zip

Figure 16. Position of astronaut at lift-off. (Reprinted with permission from A. W. Sloan: *Physiology for Students and Teachers of Physical Education.* London, Arnold, 1969.)

himself into a sleeping bag attached to the wall of the cabin. Special provision must be made for eating and drinking and disposal of excreta (see below).

Escape

Escape from a spacecraft on the launching pad or during take-off is by means of an escape rocket, which separates the command module from the service module; the subsequent fall of the command module is slowed down by parachutes. In orbit, only normal reentry procedure is practicable (Nuttall, 1971).

Space Suit

The space suit is pressurized to the equivalent of an altitude of 1,000 m and supplied with pure oxygen. Normally it need be worn only during take-off and reentry and during extravehicular activity.

Extravehicular Activity

Wearing a space suit the astronaut can emerge from the spacecraft and move about in space. He is connected to the spacecraft by a communication cable which also carries his oxygen supply. Because of its very high thermal insulation the suit must have a cooling system; but gas cooling, through the oxygen supply tube, is sufficient for short periods of extravehicular activity in space (Waligora and Horrigan, 1975). The helmet has a sun visor. Movement of the whole body is achieved by the use of a small hand-held propulsion gun.

COMPLICATIONS OF SPACE FLIGHT

Physiological Problems

Eating and Drinking

In Apollo spacecraft there was a plentiful supply of water

from the action of the fuel cells, sufficient for drinking and for topping up the cooling system, and the excess was dumped (Sauer and Calley, 1975). In a state of weightlessness water must be stored in enclosed reservoirs and drunk through a tap or from a squeeze bottle.

On the earlier flights solid food was provided in bite-sized portions or squeezed out of collapsible containers. Dried, powdered foods, vacuum packed in plastic, were rehydrated from a metered water dispenser (Vanderveen, 1971). Advances in food technology have since made possible a more varied and more appetising diet, and packages have been developed from which solid food can be eaten with a spoon (Smith et al., 1975).

Disposal of Excreta

Urine and feces are collected in bags, which may be stored or ejected from the spacecraft (Sauer and Jorgensen, 1975). For metabolic studies the 24-hour urine is measured and an aliquot sample frozen and stored; feces are dried and stored (Johnston, 1977).

Exercise

Regular exercise is necessary on flights lasting more than a day to reduce atrophy of muscle and decalcification of bone and to facilitate vasomotor readjustment on return to Earth (Hardy, 1964b). Earlier spacecraft provided tension springs, which the astronaut could stretch to perform arm and leg exercises; but more satisfactory exercise, in later spacecraft, was on a bicycle ergometer. Skylab 4 provided a simulated treadmill exercise (Thornton and Rummel, 1977).

Circadian Rhythms

In orbital flight the day-night cycle may be reduced to about 2 hours and on interplanetary travel it is abolished. The sleep rhythm is kept as close as possible to that on Earth, and the windows of the spacecraft are covered for appropriate periods

(Berry, 1971). Astronauts usually have no difficulty in sleeping, but the sleeping time tends to be reduced (Frost et al., 1977).

Medical Problems

Man cannot adapt to space but can adapt, in varying degree, to the environment of a space capsule. Further adaptation is required on return to Earth.

Disturbances of Equilibrium

Failure of adaptation to weightlessness may cause vertigo and motion sickness, a hazardous condition in a state of weightlessness because of the danger of inhaling vomitus. Antiemetic drugs are usually effective.

Minor spatial disorientation is common during the first couple of days in orbital flight and for as long as nine days after return to Earth (Homick et al., 1977).

Circulatory Disturbances

Orbiting astronauts experience a sensation of "fullness in the head" from fluid shift to the upper part of the body as a result of weightlessness (Gibson, 1977). Blood pressure is well maintained in flight, but exercise tolerance is reduced (Rummel et al., 1973); and a degree of postural hypotension is usual on return to Earth (Winter, 1977).

Metabolic Disturbances

Progressive loss of calcium, phosphorus, and nitrogen from the body, associated with decalcification of bone and wasting of muscle, seems to be inevitable in space flight, even if regular exercise is performed (Whedon et al., 1977). All the Skylab astronauts lost weight and had reduced subcutaneous fat at the end of their period in orbit (Rambaut et al., 1977). There was also some retention of sodium and loss of potassium, associated with an increased cortisol level in the blood (Leach et al., 1975).

Radiation Sickness

Since no adaptation occurs to ionizing radiation and the effects are cumulative, the danger of radiation sickness is greatest on prolonged space flight. An early sign is nausea and vomiting; the design of a space helmet must allow for this (Clark, 1964). More serious complications include destruction of reproductive tissue, of red bone marrow and, in severe cases, of intestinal epithelium. An individual who survives the acute phase is liable later to develop leukemia or some other form of malignant disease.

Tolerance of ionizing radiation can be increased temporarily by administration of cysteamine. As a precaution, some of the individual's bone marrow may be aspirated and stored for subsequent reinjection if necessary (Clark, 1964).

Hyperthermia

With earlier space suits inefficient cooling limited the performance of strenuous work. Heat stress is almost inevitable on extravehicular activities and, particularly, during reentry. Since some adaptation to heat is possible (Chapter 3), heat acclimatization has been recommended as a useful preliminary to space flight (Hardy, 1964a).

THE MOON

In the course of project Apollo, from 1969 to 1972, six pairs of American astronauts landed on the moon and explored the lunar surface (Johnston and Hull, 1975). Scientific instruments were set up, photographs taken, and samples of lunar rocks brought back to Earth. Strict quarantine was observed for the first lunar explorers and the material they brought back with them on return to Earth, but this was later abandoned when no disease-producing organisms were found (Johnston et al., 1975).

Lunar Environment

Like outer space, the lunar environment is incapable of supporting life as we know it. There is no surface water and no atmosphere, except for traces of helium and argon (Marvin, 1973). Consequently there is no protection from solar and cosmic radiation. Temperatures are extreme; during the day (about 14 terrestrial days) the surface temperature rises to about 93°C, and during the night it falls as low as −240°C (Simmons, 1971). The force of gravity on the moon's surface is approximately one sixth that on the surface of Earth.

Man on the Moon

On the moon man must wear his pressure suit, which also provides thermal insulation, and he must carry his own oxygen supply (Fig. 17). The extravehicular mobility unit for use on the moon comprises a pressure suit (maintained at about 190 mm Hg) with instrumentation and communications equipment, a liquid cooling system, and a collecting bag for urine. The portable life support system, carried on the back, provides oxygen for pressure and for respiration, absorbs carbon dioxide, and drives the liquid cooling system. An emergency oxygen purge system, manually operated, provides oxygen in the event of failure of the main system, but it has never had to be used (Carson et al., 1975). The space helmet contains a high-energy food bar and a drinking device (Smith et al., 1975). It has a lunar visor, which gives protection against light, heat, and micrometeorites (Carson et al., 1975). Pressure gloves and lunar boots complete the extravehicular mobility unit.

Low gravity and the bulky space suit make ordinary walking difficult, and progress is easier with large steps or bounds. Though covered with fine dust, the surface is sufficiently firm to carry a wheeled vehicle, i.e. the Lunar Rover carried on the later Apollo flights (Carson et al., 1975).

Possibly because of the unusual demands on muscles and the restrictive effects of a space suit, movements on the moon are hard work, as evidenced by the high heart rates of lunar explorers monitored by telemetry at the base on Earth. Thermal stress is avoided by the liquid cooling system in the space suit, which can dissipate heat at the rate of 500 kcal/hr; the actual work load seldom exceeds 300 kcal/hr (Waligora and Horrigan, 1977). During a prolonged stay on the moon some adaptation

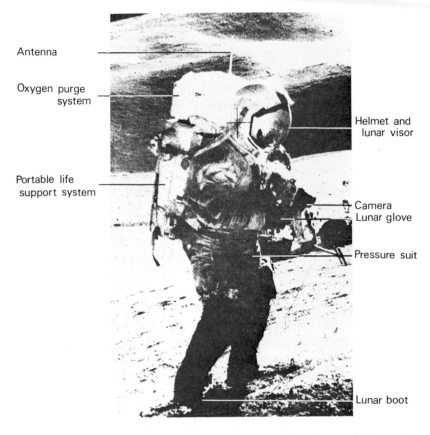

Figure 17. Man on the moon; Apollo 15 extravehicular mobility unit. (Courtesy of the National Aeronautics and Space Administration.)

to the low gravity might be expected, but man must continue to provide his own habitable immediate environment.

OTHER PLANETS

Modern technology has made it possible to explore other planets with unmanned spacecraft. When the problems of sustaining life in a space capsule for many months have been solved, man should be able to visit Mars and possibly Venus, though the climate of Venus is even more inhospitable (see below). The planets further away have even more extreme environmental conditions, for instance the intense heat of Mercury and the very high gravitational force of Jupiter.

Mars

The environment on Mars has much in common with that on the moon. The surface is solid, with a layer of fine dust strewn with boulders (Gore, 1977). The force of gravity is about two-fifths that on the surface of Earth. There is no surface water, but water ice as well as carbon dioxide ice is found at the poles and probably as permafrost elsewhere. The Martian day is about 24 hours and the year 687 days; since the axis of rotation, like that of Earth, is inclined about 24° to the plane of the orbit there are seasonal changes. Cold is extreme; the diurnal range of temperature in summer is −86°C to −31°C (Gore, 1977).

Atmosphere

The atmosphere of Mars is thin, exerting only about one-two hundredth Earth's atmospheric pressure at the surface (Nier et al., 1976). It is predominantly carbon dioxide, with some argon and nitrogen, very little water vapor, and practically no oxygen (Pollack, 1975). Dust storms are frequent.

Venus

Venus has a solid surface, details of which are hidden from Earth by an atmosphere of dense clouds. The force of gravity is similar to that of Earth. The planet rotates on its axis in 243 days, a little longer than the period of its orbit around the sun, but the "greenhouse effect" of the very dense atmosphere keeps the surface temperature fairly constant at about 480°C (Weaver, 1975).

Atmosphere

Atmospheric pressure on the surface of Venus is about ninety times that on the surface of Earth. The atmosphere is mainly carbon dioxide, with little water and no oxygen but with some sulphuric acid in the upper atmosphere (Young and Young, 1975). Exploring Venus would be "like wandering around the depths of a boiling ocean in a submarine" (Shelton, 1972).

Prolonged Space Flight

The journey to Mars, with present-day equipment, takes about 8 months and the journey to Venus about 4 months; it would take about 1 1/2 years to reach Jupiter (Shelton, 1972). Apart from the as yet unknown psychological problems of isolation in spacecraft for such long periods, and the cumulative effect of cosmic radiation (see above), the main problems will be the maintenance of a habitable environment and provision of food and water. There is also the danger that weightlessness lasting more than 6 months may cause irreversible damage to bone and to muscle (Winter, 1977).

Recycling

Some economy of payload can be achieved if dehydrated food is carried and reconstituted with water recovered from urine, and from water vapor in expired air, and evaporated from the skin (Welch, 1963; Brobeck, 1964).

Oxygen may be renewed and carbon dioxide disposal supplemented by photosynthesis by green plants exposed to sunlight; this may also supplement the food supply. At present the most promising regenerative system is provided by the alga *Chlorella,* but it requires a considerable mass of culture solution and more combined nitrogen than could be recovered from the astronauts' urine (Adams, 1975). So far the food prepared from it is not very palatable, and palatability is of major importance in the food provided for prolonged space travel (Brobeck, 1964).

Chlorella also forms some carbon monoxide. Removal of carbon monoxide and other toxic gases can be achieved by the techniques employed in a nuclear submarine (Chapter 5).

Depressed Metabolism

Some form of "suspended animation," due to depression of

metabolism by appropriate drugs, would conserve energy and reduce the demands on food and oxygen; but at present this is still in the realm of science fiction.

METRIC UNITS AND
THEIR EQUIVALENTS

OLDER units are being replaced by metric and some older metric units by those of the International System (Système International, SI). Some units have acquired a new name, e.g. millimeters of mercury = torr; cycles per second = hertz; degree centigrade = degree Celsius. Conversion factors from old to new units are given in the following tables and there is a conversion chart for measurements of temperature.

METRIC UNITS

Quantity	Old unit	Metric unit	Equivalent
length	foot (ft)	meter (m)	3.281 ft
volume	pint (USA)	liter (l)	2.114 pints
	" (UK)	"	1.760 pints
mass	pound (lb)	kilogram (kg)	2.205 lb
velocity	feet per second (ft/sec)	meters per second (m/sec)	3.281 ft/sec
"	miles per hour (mph)	kilometers per hour (kph)	0.621 mph
force	poundal (pdl)	dyne (dyn)	7.233×10^{-5} pdl
pressure	pounds per square inch (psi)	dynes per square centimeter (dyn/cm²)	1.450×10^{-5} psi
work, energy	foot pound (ft lb)	erg (erg)	7.373×10^{-8} ft lb
heat energy	British thermal unit (BTU)	calorie (cal)	3.968×10^{-3} BTU
power	horse power (HP)	calories per second (cal/sec)	5.613×10^{-3} HP
temperature	degree Fahrenheit (°F)	degree Celsius (°C)	1.800 °F*

*Since freezing point on the Fahrenheit scale is 32°C, this figure must be subtracted from the Fahrenheit reading before it is converted to °C. See conversion chart.

SI UNITS

Quantity	Old metric unit	SI unit	Equivalent
force	dyne (dyn)	newton (N)	10^5 dyn
pressure	dynes per square centimeter (dyn/cm²)	pascal (Pa)	10 dyn/cm²
energy	calorie (cal)	joule (J)	4.186 cal
power	calorie per hour (cal/hr)	watt (W)	1.507×10^4 cal/hr

PHYSICAL CONSTANTS

Acceleration due to gravity (g) = 981 cm/sec² (32 ft/sec²)
Atmospheric pressure at sea level (atm) = 1.013×10^5 Pa
(14.7 lb/in², 760 mm Hg, 10 m sea water)

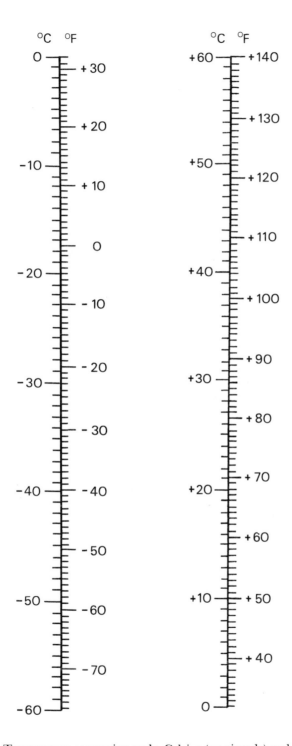

Figure 18. Temperature conversion scale, Celsius (centigrade) and Fahrenheit.

BIBLIOGRAPHY

Adams, C.C.: Nutritional aspects of space flight. In Bourne, G.H. (Ed.): *Medical and Biological Problems of Space Flight.* New York, Academic Press, 1975, pp. 237-244.

Adams, T. and Heberling, E.J.: Human physiological responses to a standardized cold stress as modified by physical fitness. *J Appl Physiol, 13:*226-230, 1958.

Andersen, H.T.: Physiological adaptations in diving vertebrates. *Physiol Rev, 46:*212-243, 1966.

Anthonissen, N.R.: Respiratory system in diving. In Strauss, R.H. (Ed.): *Diving Medicine.* New York, Grune and Stratton, 1976, pp. 35-48.

Armstrong, H.G.: *Principles and Practice of Aviation Medicine,* 2nd ed. Baltimore, Williams and Wilkins, 1943.

Asmussen, E.: The regulation of respiration. In Cunningham, D.J.C. and Lloyd, B.B. (Eds.): *The Regulation of Human Respiration.* Oxford, Blackwell, 1963, pp. 59-70.

Åstrand, P.-O. and Rodahl, K.: *Textbook of Work Physiology.* New York, McGraw-Hill, 1970, p. 71.

Bachrach, A.J.: A short history of man in the sea. In Bennett, P.B. and Elliot, D.H. (Eds.): *The Physiology and Medicine of Diving,* 2nd ed. London, Baillière Tindall, 1975, pp. 1-10.

Bailey, J.V., Hoffman, R.A. and English, R.A.: Radiological protection and medical dosimetry for the Skylab crewmen. In Johnston, R.S. and Dietlein, L.F. (Eds.): *Biomedical Results from Skylab.* Washington, D.C., NASA, 1977, pp. 64-72.

Baker, P.T.: The adaptive fitness of high altitude populations. In Baker, P.T. (Ed.): *The Biology of High Altitude Populations.* Cambridge, University Press, 1978, pp. 317-350.

Baker, P.T., Pawson, I.G. and Weitz, C.A.: High altitude studies in Nepal. In Collins, K.J. and Weiner, J.S. (Eds.): *Human Adaptability.* London, Taylor and Francis, 1977, pp. 285-287.

Balke, B.: Human tolerances. In Brown, J.H.V. (Ed.): *Physiology of Man in Space.* New York, Academic Press, 1963, pp. 149-171.

Barnard, C.: A human cardiac transplant: an interim report of a successful operation performed at Groote Schuur Hospital, Cape Town. *S Afr Med J, 41:*1271-1274, 1967.

Bazett, H.C.: The regulation of body temperatures. In Newburgh, L.H. (Ed.): *Physiology of Heat Regulation and the Science of Clothing.* New

York, Hafner, 1949, pp. 109-192.

Bedford, T.: *Environmental Warmth and its Measurement.* Medical Research Council War Memorandum No. 17. London, H.M. Stationery Office, 1946, p. 24.

Beighton, P.: Fluid balance in the Sahara. *Nature (Lond), 233*:275-277, 1971.

Belding, H.S.: Some principles of acclimatization to heat. In Yousef, M.K. and Horvath, S.M. (Eds.): *Physiological Adaptations: Desert and Mountain.* New York, Academic Press, 1972, pp. 9-21.

Bell, G. H., Emslie-Smith, D. and Paterson, J.R.: *Textbook of Physiology and Biochemistry,* 9th ed. New York, Churchill Livingstone, 1976.

Bennett, P.B.: Narcotic action of inert gases. In Edholm, O.G. and Bacharach, A.L. (Eds.): *The Physiology of Human Survival.* New York, Academic Press, 1965, pp. 164-182.

Bennett, P.B.: Performance impairment in deep diving due to nitrogen, helium, neon and oxygen. In Lambertsen, C.J. (Ed.): *Underwater Physiology.* Baltimore, Williams and Wilkins, 1967, pp. 327-340.

Bennett, P.B.: The high pressure nervous syndrome. In Bennett, P.B. and Elliott, D.H. (Eds.): *The Physiology and Medicine of Diving,* 2nd ed. London, Baillière Tindall, 1975, pp. 248-263.

Bennett, P.B.: The physiology of nitrogen narcosis and the high pressure nervous syndrome. In Strauss, R.H. (Ed.): *Diving Medicine.* New York, Grune and Stratton, 1976, pp. 157-181.

Benzinger, T.H.: Heat regulation: homeostasis of central temperature in man. *Physiol Rev, 49*:671-749, 1969.

Bernard, C.: *Leçons sur les Phénomènes de la vie communs aux Animaux et aux Végétaux.* Paris, Baillière, 1878.

Berry, C.A.: Medical experience in manned space flight. In Randel, H.W. (Ed.): *Aerospace Medicine,* 2nd ed. Baltimore, Williams and Wilkins, 1971, pp. 697-718.

Berry, C.A.: Medical legacy of Apollo. *Aerospace Med, 45*:1046-1057, 1974.

Bert, P.: *La Pression Barométrique.* English translation by Hitchcock, M.A. and Hitchcock, F.A.: Columbus, Ohio, College Book Co., 1943, p. 23.

Bligh, J. and Johnson, K.G.: Glossary of terms for thermal physiology. *J Appl Physiol, 35*:941-961, 1973.

Bodey, A.S.: The role of catecholamines in human acclimatization to cold: a study of 24 men at Casey, Antarctica. In Edholm, O.G. and Gunderson, E.K.E. (Eds.): *Polar Human Physiology.* London, Heinemann, 1973, pp. 141-149.

Bond, G.F.: Medical problems of multiday saturation diving in open water. In Lambertsen, C.J. (Ed.): *Underwater Physiology.* Baltimore, Williams and Wilkins, 1967, pp. 81-88.

Bornmann, R.C.: Decompression after saturation diving. In Lambertsen, C.J. (Ed.): *Underwater Physiology.* Baltimore, Williams and Wilkins, 1967, pp. 109-121.

Bourdillon, T.D.: The use of oxygen apparatus by acclimatized men. *Proc Roy Soc Ser B, 143*:24-32, 1954.

Boycott, A.E. and Damant, G.C.C.: Experiments on the influence of fatness on susceptibility to caisson disease. *J Hyg England*, 8:445-456, 1908.

Boycott, A.E., Damant, G.C.C. and Haldane, J.S.: The prevention of compressed air illness. *J Hyg England*, 8:342-443, 1908.

Brady, J.C., Hughes, D.F., Samonski, F.H., Jr., Young, R.W. and Browne, D.M.: In Johnston, R.S., Dietlein, L.F. and Berry, C.A. (Eds.): *Biomedical Results of Apollo*. Washington, D.C., NASA, 1975, pp. 517-543.

Brobeck, J.R.: Food requirements in space. In Hardy, J.D. (Ed.): *Physiological Problems in Space Exploration*. Springfield, Thomas, 1964, pp. 134-151.

Brooks, R.E., Natani, K., Shurley, J.T., Pierce, C.M. and Joern, A.T.: An antarctic sleep and dream laboratory. In Edholm, O.G. and Gunderson, E.K.E. (Eds.): *Polar Human Biology*. London, Heinemann, 1973, pp. 322-341.

Brotherhood, J.R.: Studies on energy expenditure in the Antarctic. In Edholm, O.G. and Gunderson, E.K.E. (Eds.): *Polar Human Biology*. London, Heinemann, 1973, pp. 182-192.

Brown, A.H.: Dehydration exhaustion. In Adolph, E.F. (Ed.): *Physiology of Man in the Desert*. New York, Interscience Publishers, 1974, pp. 208-225.

Brown, H.H.S.: The pressure cabin. In Gillies, J.A. (Ed.): *A Textbook of Aviation Physiology*. Oxford, Pergamon, 1965, pp. 152-186.

Budd, G.M.: Australian physiological research in the Antarctic and the Subantarctic, with special reference to thermal stress and acclimatization. In Edholm, O.G. and Gunderson, E.K.E. (Eds.): *Polar Human Biology*. London, Heinemann, 1973, pp. 15-40.

Bullard, R.W.: Physiological problems of space travel. *Ann Rev Physiol*, 34:205-234, 1972.

Burton, A.C. and Edholm, O.G.: *Man in a Cold Environment*. London, Arnold, 1955.

Buskirk, E.R.: Work capacity of high-altitude natives. In Baker, P.T. (Ed.): *The Biology of High Altitude Peoples*. Cambridge, University Press, 1978, pp. 173-187.

Cabanac, M.: Temperature regulation. *Ann Rev Physiol*, 37:415-439, 1975.

Cannon, W.B.: *The Wisdom of the Body*, 2nd ed. New York, Norton, 1939, p. 24.

Carlson, L.D. and Hsieh, A.S.L.: Cold. In Edholm, O.G. and Bacharach, A.L. (Eds.): *The Physiology of Human Survival*. New York, Academic Press, 1965, pp. 15-52.

Carson, M.A., Rouen, M.N., Lutz, C.C. and McBarron, J.W. II: Extravehicular mobility unit. In Johnston, R.S., Dietlein, L.F. and Berry, C.A. (Eds.): *Biomedical Results of Apollo*. Washington, D.C., NASA, 1975, pp. 545-569.

Cerretelli, P.: High altitude physiology and medicine. In Monzino, G.: *La Spedizione Italiana all' Everest*. Verona, Monzino, 1976, pp. 213-247.

Christy, R.L.: Effects of radial, angular, and transverse acceleration. In Randel, H.W. (Ed.): *Aerospace Medicine*, 2nd ed. Baltimore, Williams and Wilkins, 1971, pp. 167-197.

Chiodi, H.: Respiratory adaptation to high altitude. In Cunningham, D.J.C. and Lloyd, B.B. (Eds.): *The Regulation of Human Respiration*. Oxford, Blackwell, 1963.

Clark, C.: High energy radiations. In Hardy, J.D. (Ed.): *Physiological Problems in Space Exploration*. Springfield, Thomas, 1964, pp. 47-99.

Clegg, E.J.: Fertility and early growth. In Baker, P.T. (Ed.): *The Biology of High Altitude Peoples*. Cambridge, University Press, 1978, pp. 65-115.

Clegg, E.J., Harrison, G.A. and Baker, P.T.: The impact of high altitudes on human population. *Hum Biol, 42:*486-518, 1970.

Collins, K.J.: Endocrine control of salt and water in hot conditions. *Fed Proc, 22:*716-720, 1963.

Comroe, J.H.: *Physiology of Respiration*, 2nd ed. Chicago, Year Book Medical Publishers, 1974.

Consolazio, C.F.: The energy requirements of men living under extreme environmental conditions. *Wld Rev Nutr Diet, 4:*57-77, 1963.

Consolazio, C.F., Johnson, H.L. and Krzyicki, H.J.: Body fluids, body composition and metabolic aspects of high-altitude adaptation. In Yousef, M.K., Horvath, S.M. and Bullard, R.W. (Eds.): *Physiological Adapatations: Desert and Mountain*. New York, Academic Press, 1972, pp. 227-241.

Consolazio, C.F., Johnson, R.E. and Pecora, L.J.: *Physiological Measurements of Metabolic Functions in Man*. New York, McGraw-Hill, 1963, pp. 410-413.

Cooper, K.E.: Hypothermia. In Strauss, R.H. (Ed.): *Diving Medicine*. New York, Grune and Stratton, 1976, pp. 211-226.

Cousteau, J.Y.: *The Silent World*. London: Reprint Society, 1953.

Das, S.K. and Saha, H.: The respiratory metabolism of the Sherpas (Hill-people) during climbing. *Indian J Med Res, 55:*579-583, 1967.

Dickson, J.G., Jr. and MacInnis, J.B.: Confluence of physiological environmental and engineering factors in prolonged diving at extreme depths. In Lambertson, C.J. (Ed.): *Underwater Physiology*. Baltimore, Williams and Wilkins, 1967, pp. 89-97.

Dietrick, J.E., Whedon, G.D. and Shorr, E.: Effects of immobilization upon various metabolic and physiologic functions of normal men. *Am J Med, 4:*3-36, 1948.

Donayre, J., Guerra-Garcia, R., Moncloa, F. and Sobrevilla, L.A.: Endocrine studies at high altitude IV. Changes in the semen of men. *J Reprod Fertil, 16:*55-58, 1968.

Dougherty, J.H., Jr.: Use of H_2 as an inert gas during diving: pulmonary function during H_2-O_2 breathing at 7.06 ATA. *Aviat Space Environ Med, 47:*618-626, 1976.

Edholm, O.G.: The effect in man of acclimatization to heat on water intake, sweat rate and water balance. In Itoh, S., Ogata, K. and Yoshimura, H.

(Eds.): *Advances in Climatic Physiology.* Tokyo, Igaku Shoin, 1972, pp. 144-155.

Edholm, O.G., Hackett, A.J., Sakkar, M.Y. and Weiner, J.S.: The assessment of levels of heat tolerance. *J Physiol, 268:*9p-10p, 1977.

Edholm, O.G. and Lewis, H.E.: Terrestial animals in cold: man in polar regions. In Dill, D.B., Adolph, E.F. and Wilber, C.G. (Eds.): *Handbook of Physiology,* Section 4. Washington, D.C., Am Physiol Soc, 1964, pp. 259-282.

Edmonds, C.: First aid and emergency medical treatment. In Strauss, R.H. (Ed.): *Diving Medicine.* New York, Grune and Stratton, 1976, pp. 309-316.

Egstrom, G.H.: Diving equipment. In Strauss, R.H. (Ed.): *Diving Medicine.* New York, Grune and Stratton, 1976, pp. 23-24.

Elliott, D.H. and Hallenbeck, J.M.: The pathophysiology of decompression sickness. In Bennett, P.B. and Elliott, D.H. (Eds.): *The Physiology and Medicine of Diving,* 2nd ed. London, Baillière Tindall, 1975, pp. 435-455.

Ernsting, J.: The physiological requirements of aircraft oxygen systems. In Gillies, J.A. (Ed.): *A Textbook of Aviation Physiology.* Oxford, Pergamon, 1965, pp. 303-342.

Ernsting, J. and Stewart, W.K.: Oxygen deprivation at reduced barometric pressure. In Gillies, J.A. (Ed.): *A Textbook of Aviation Physiology.* Oxford, Pergamon, 1965, pp. 209-213.

Fagan, P., McKenzie, B. and Edmonds, C.: Sinus barotrauma in divers. *Ann Otol Rhinol Laryngol, 85:*61-64, 1976.

Fanger, P.O.: Assessment of man's thermal comfort in practice. *Brit J Indust Med, 30:*313-324, 1973.

Farmer, J.C. and Thomas, W.G.: Auditory and vestibular function in diving. In Bennett, P.B. and Elliott, D.H. (Eds.): *The Physiology and Medicine of Diving,* 2nd ed. London, Baillière Tindall, 1975, pp. 524-544.

Fox, R.H.: Heat. In Edholm, O.G. and Bacharach, A.L. (Eds.): The Physiology of Human Survival. New York, Academic Press, 1965, pp. 53-79.

Fox, R.H.: Temperature regulation with special reference to man. In Linden, R.J. (Ed.): *Recent Advances in Physiology.* Edinburgh, Churchill Livingstone, 1974, pp. 340-405.

Fox, R.H.: Heat tolerance studies of different ethnic groups. In Collins, K.J. and Weiner, J.S. (Eds.): *Human Adaptability: a History and Compendium of Research.* London, Taylor and Francis, 1977, pp. 277-278.

Fox, R.H., MacDonald, I.C. and Woodward, P.M.: A hypothermia survey kit. *J Physiol, 231:*4p-7p, 1973.

Frisancho, A.R. and Baker, P.T.: Altitude and growth: a study of the patterns of physical growth of a high altitude Peruvian Quechua population. *Am J Phys Anthrop, 32:*279-292, 1970.

Frost, J.D., Shumate, W.H., Salamy, J.G. and Booker, C.R.: Sleep monitoring on skylab. In Johnston, R.A. and Dietlein, R.F. (Eds.): *Biomedical Results from Skylab.* Washington, D.C., NASA, 1977, pp. 113-126.

Fryer, D.I.: Failure of the pressure cabin. In Gillies, J.A. (Ed.): *A Textbook of Aviation Physiology.* Oxford, Pergamon, 1965, pp. 187-206.

Fryer, D.I. and Roxburgh, H.L.: Decompression sickness. In Gillies, J.A. (Ed.): *A Textbook of Aviation Physiology.* Oxford, Pergamon, 1965, pp. 122-151.

Fürer-Haimendorf, C.V.: *The Sherpas of Nepal.* London, Murray, 1964, p. 2.

Gagge, A.-P., Winslow, C.-E.A. and Herrington, L.P.: The influence of clothing on the physiological reactions of the human body to varying environmental temperatures. *Am J Physiol, 124:*30-50, 1938.

Ganong, W.F.: *Review of Medical Physiology,* 8th ed. Los Altos, Lange, 1977.

Gelineo, S.: Organ systems in adaptation: the temperature regulating system. In Dill, D.B., Adolph, E.F. and Wilber, C.G. (Eds.): *Handbook of Physiology,* Section 4. Washington, D.C., Am Physiol Soc, 1964, pp. 259-282.

Gibson, E.G.: Skylab 4 crew observations. In Johnston, R.S. and Dietlein, L.F. (Eds.): *Biomedical Results from Skylab.* Washington, D.C., NASA, 1977, pp. 22-26.

Glaister, J. and Rentoul, E.: *Medical Jurisprudence and Toxicology,* 12th ed. Edinburgh, Livingstone, 1966, p. 149.

Glaser, E.M.: Immersion and survival in cold water. *Nature (Lond), 166:*1068, 1950.

Godin, G. and Shephard, R.J.: Activity patterns of the Canadian Eskimo. In Edholm, O.G. and Gunderson, E.K.E. (Eds.): *Polar Human Biology.* London, Heinemann, 1973, pp. 193-215.

Goldsmith, R.: Evidence of acclimatization to cold obtained from clothing records. *J Physiol, 148:*79p-80p.

Gore, R.: Sifting for life in the sands of Mars. *Nat Geograph Mag, 151:*8-31, 1977.

Gosselin, R.E.: Rates of sweating in the desert. In Adolph, E.F. (Ed.): *Physiology of Man in the Desert.* New York, Interscience Publishers, 1947, pp. 44-76.

Guignard, J.C.: Noise. In Gillies, J.A. (Ed.): *A Textbook of Aviation Physiology.* Oxford, Pergamon, 1965, pp. 895-967.

Guyton, A.C.: *Textbook of Medical Physiology,* 5th ed. Philadelphia, Saunders, 1976.

Haldane, J.S. and Priestley, J.G.: The regulation of the lung ventilation. *J Physiol, 32:*225-266, 1905.

Halstead, B.W.: Hazardous marine life. In Strauss, R.H. (Ed.): *Diving Medicine.* New York, Grune and Stratton, 1976, pp. 227-256.

Hammel, H.T.: Effect of race on response to cold. *Fed Proc, 22:*795-800, 1963.

Hammel, H.T.: Terrestrial animals in cold: recent studies of primitive man. In Dill, D.B., Adolph, E.F. and Wilber, C.G. (Eds.): *Handbook of Physiology,* Part 4. Washington, D.C., Am Physiol Soc, 1964, pp. 413-434.

Hardy, J.D.: Physiology of temperature regulation. *Physiol Rev, 41:*521-606, 1961.

Hardy, J.D.: Temperature problems in space travel. In Hardy, J.D. (Ed.): *Physiological Problems in Space Exploration.* Springfield, Thomas, 1964, pp. 3-46.

Hardy, J.D.: Weightlessness and sub-gravity problems. In Hardy, J.D. (Ed.): *Physiological Problems in Space Exploration.* Springfield, Thomas, 1964, pp. 196-208.

Harrison, G.A., Küchemann, C.F., Moore, M.A.S., Boyce, A.J., Baju, T., Mourant, A.E., Godber, M.J., Glasgow, B.G., Kopéc, A.C., Tills, D. and Clegg, E.J.: The effects of altitudinal variation in Ethiopian populations. *Phil Trans Roy Soc Ser B, 256:*147-182, 1969.

Harrison, G.A., Weiner, J.S., Tanner, J.M. and Barnicot, N.A.: *Human Biology: an Introduction to Human Evolution, Variation and Growth.* Oxford, University Press, 1964, p. 204.

Hawkins, W.R. and Zieglschmid, J.F.: Clinical aspects of crew health. In Johnston, R.S., Dietlein, L.F. and Berry, C.A. (Eds.): *Biomedical Aspects of Apollo.* Washington, D.C., NASA, pp. 43-81.

Hayward, J.S., Eckerson, J.D. and Collis, M.L.: Thermal balance and survival time prediction for man in cold water. *Can J Physiol Pharmacol, 53:*21-32, 1975.

Heath, D.A. and Williams, D.R.: *Man at High Altitude: the Pathophysiology of Acclimatization and Adaptation.* Edinburgh, Churchill Livingstone, 1977.

Hempleman, H.V.: Decompression theory: British practice. In Bennett, P.B. and Elliott, D.H. (Eds.): *The Physiology and Medicine of Diving,* 2nd ed. London, Baillière Tindall, 1975, pp. 331-347.

Herrington, L.P.: The range of physiological response to climatic heat and cold. In Newburgh, L.H. (Ed.): *Physiology of Heat Regulation and the Science of Clothing.* New York, Hafner, 1949, pp. 262-276.

Hildes, J.A.: Comparison of coastal Eskimos and Kalahari Bushmen. *Fed Proc, 22:*843-845, 1963.

Hill, L.: *Caisson Sickness and the Physiology of Work in Compressed Air.* London, Arnold, 1912, p. 164.

Hock, R.J.: The physiology of high altitude. *Sci Amer, 222(2):*53-62, 1970.

Hock, R.J. and Covino, B.G.: Hypothermia. *Sci Amer, 198(3):*104-114, 1958.

Homick, J.L., Reschke, M.F. and Miller, E.F., II: The effects of prolonged exposure to weightlessness on postural equilibrium. In Johnston, R.A. and Dietlein, L.F. (Eds.): *Biomedical Results from Skylab.* Washington, D.C., NASA, 1977, pp. 104-112.

Hong, S.K.: Comparison of diving and nondiving women of Korea. *Fed Proc, 22:*831-833, 1963.

Hong, S.K.: The physiology of breath-hold diving. In Strauss, R.H. (Ed.): *Diving Medicine.* New York, Grune and Stratton, 1976, pp. 269-286.

Hong, S.K. and Rahn, H.: The diving women of Korea and Japan. In Vander, A.J. (Ed.): *Human Physiology and the Environment in Health and*

Disease. San Francisco, Freeman, 1976, pp. 92-101.

Hong, S.K., Rahn, H., Kang, D.H., Song, S.H. and Kang, B.S.: Diving pattern, lung volumes, and alveolar gas of the Korean diving woman (ama). *J Appl Physiol, 18:*457-465, 1963.

Houghten, F.C., Teague, W.W., Miller, W.E. and Yant, W.P.: Thermal exchanges between the human body and its atmospheric environment. *Am J Physiol, 88:*386-406, 1929.

Houghten, F.C. and Yagloglou, C.P.: Determining equal comfort lines. *J Am Soc Heat-Vent Engnrs, 29:*165-176, 1923a.

Houghten, F.C. and Yagloglou, C.P.: Determination of the comfort zone. *J Am Soc Heat-Vent Engnrs, 29:*515-532, 1923b.

Howard, P.: High and low gravitational forces. In Edholm, O.G. and Bacharach, A.L. (Eds.): *The Physiology of Human Survival.* New York, Academic Press, 1965a, pp. 183-206.

Howard, P.: The physiology of positive acceleration. In Gillies, J.A. (Ed.): *A Textbook of Aviation Physiology.* Oxford, Pergamon, 1965b, pp. 551-687.

Howard, P.: The physiology of negative acceleration. In Gillies, J.A. (Ed.): *A Textbook of Aviation Physiology.* Oxford, Pergamon, 1965c, pp. 688-716.

Howard, P.: The physiology of transverse acceleration. In Gillies, J.A. (Ed.): *A Textbook of Aviation Physiology.* Oxford, Pergamon, 1965d, pp. 717-745.

Howell, J.B.: The sunlight factor in aging and skin cancer. *Arch Dermatol, 82:*865-869.

Hunt, J.: *The Ascent of Everest.* London, Hodder and Stoughton, 1953, p. 205.

Hurtado, A.: Natural acclimatization to high altitudes. In Cunningham, D.J.C. and Lloyd, B.B. (Eds.): *The Regulation of Human Respiration.* Oxford, Blackwell, 1963, pp. 71-82.

Hurtado, H.: Animals in high altitudes: resident man. In Dill, D.B., Adolph, E.F. and Wilber, C.G. (Eds.): *Handbook of Physiology,* Section 4. Washington, D.C., Am Physiol Soc, 1964, pp. 843-860.

Hurtado, A.: The influence of high altitude on physiology. In Porter, R. and Knight, J. (Eds.): *High Altitude Physiology: Cardiac and Respiratory Aspects.* Edinburgh, Churchill Livingstone, 1971, pp. 3-8.

Ingram, D.L.: Physiological reactions to heat in man. In Garlick, J.P. and Keay, R.W.J. (Eds.): *Human Ecology in the Tropics,* 2nd ed. London, Taylor and Francis, 1977, pp. 95-112.

Itoh, S.: *Physiology of Cold-adapted Man.* Supporo, Hokkaido University, 1974.

Jenkins, T. and Nurse, G.T.: Biomedical studies in the desert dwelling hunter-gatherers of Southern Africa. *Prog Med Genet (NS), 1:*213-281.

Johnston, R.S.: Skylab medical program overview. In Johnston, R.S. and Dietlein, L.F. (Eds.): *Biomedical Results from Skylab.* Washington, D.C., NASA, 1977, pp. 3-19.

Johnston, R.S. and Hull, W.E.: Apollo missions. In Johnston, R.S., Dietlein, L.F. and Berry, C.A. (Eds.): *Biomedical Results of Apollo.* Washington, D.C., NASA, 1975, pp. 9-40.

Johnston, R.S., Mason, J.A., Wooley, B.C., McCollum, G.W. and Mieszuk, B.J.: The lunar quarantine program. In Johnston, R.S., Dietlein, L.F. and Berry, C.A. (Eds.): *Biomedical Results of Apollo.* Washington, D.C., NASA, pp. 407-424.

Jones, J. and Jones, G.M.: Aerodynamic forces and their effects upon man. In Gillies, J.A. (Ed.): *A Textbook of Aviation Physiology.* Oxford, Pergamon, 1965, pp. 55-94.

Kaplan, B.H.: Method of determining spinal alignment and level of probable fracture during static evaluation of ejection seats. *Aerospace Med, 45:*942-944, 1974.

Karstens, A.I and Welch, B.E.: Spacecraft atmospheres. In Randell, H.W. (Ed.): *Aerospace Medicine.* Baltimore, Williams and Wilkins, 1971, pp. 664-683.

Keatinge, W.R.: *Survival in Cold Water.* Oxford, Blackwell, 1969.

Kellog, R.H.: The role of CO_2 in altitude acclimatization. In Cunningham, D.J.C. and Lloyd, B.B. (Eds.): *The Regulation of Human Respiration.* Oxford, Blackwell, 1963, pp. 379-395.

Kenntner, G.: Gebräuche und Leistungsfähigkeiten im Tragen von Lasten bei Bewohnern des südlichen Himalaya. *J Morphol Anthropol, 61:*125-169, 1969.

Kerslake, D.M.: *The Stress of Hot Environments.* Cambridge, University Press, 1972.

Kerslake, D.M., Nelms, J.D. and Billingham, J.: Thermal stress in aviation. In Gillies, J.A. (Ed.): *A Textbook of Aviation Physiology.* Oxford, Pergamon, 1965, pp. 441-478.

Keys, A.: The physiology of life at high altitudes: the international high altitude expedition to Chile, 1935. *Sci Monthly, 43:*289-312, 1936.

Kidd, D.J. and Elliott, D.H.: Decompression disorders in divers. In Bennett, P.B. and Elliott, D.H. (Eds.): *The Physiology and Medicine of Diving,* 2nd ed. London, Baillière Tindall, 1975, pp. 471-495.

Kindwall, E.P.: A short history of diving and diving medicine. In Strauss, R.H. (Ed.): *Diving Medicine.* New York, Grune and Stratton, 1976a, pp. 1-12.

Kindwall, E.P.: Hyperbaric and ancillary treatment of decompression sickness, air embolism, and related disorders. In Strauss, R.H. (Ed.): *Diving Medicine.* New York, Grune and Stratton, 1976b, pp. 83-96.

Kobayasi, S.: Studies on the physiological adaptability of underwater workers. In Collins, K.J. and Weiner, J.S. (Eds.): *Human Adaptability: a History and Compendium of Research.* London, Taylor and Francis, 1977, pp. 171-172.

Krefft, S.: Cardiac injuries resulting from ejection. *Aerospace Med, 45:*948-953, 1974.

Kryter, K.D.: *The Effects of Noise on Man.* New York, Academic Press, 1970.

Ladell, W.S.: Terrestrial animals in humid heat: man. In Dill, D.B., Adolph, E.F. and Wilber, C.G. (Eds.): *Handbook of Physiology*, Section 4. Washington, D.C., Am Physiol Soc, 1964, pp. 259-282.

Lahiri, S.: Genetic aspects of the blunted chemoreflex ventilatory response to hypoxia on high altitude adaptation. In Porter, R. and Knight, J. (Eds.): *High Altitude Physiology: Cardiac and Respiratory Aspects*. Edinburgh, Churchill Livingstone, 1971, pp. 103-111.

Leach, C.S., Alexander, W.C. and Johnson, P.C.: Endocrine, electrolyte and fluid volume changes, associated with Apollo missions. In Johnston, R.S., Dietlein, L.F. and Berry, C.A. (Eds.): *Biomedical Results of Apollo*. Washington, D.C., NASA, 1975, pp. 163-184.

LeBlanc, J.: Subcutaneous fat and skin temperature. *Can J Biochem Physiol, 32*:354-358, 1954.

LeBlanc, J.: Incidence and meaning of acclimatization to cold in man. *J Appl Physiol, 9*:395-398, 1956.

LeBlanc, J.: *Man in the Cold*. Springfield, Thomas, 1975.

Lee, D.H.K.: Terrestrial animals in dry heat: man in the desert. In Dill, D.B., Adolph, E.F. and Wilber, C.G. (Eds.): *Handbook of Physiology*, Section 4. Washington, D.C., Am Physiol Soc, 1964, pp. 551-582.

Leith, I.: Serum thyroxine and triiodothyronine responses to cold in man. In Edholm, O.G. and Gunderson, E.K.E. (Eds.): *Human Polar Biology*. London, Heinemann, 1973, pp. 150-153.

Leithead, C.S.: Comfort and efficiency in the tropics. *Lancet, i:*1281-1282, 1961.

Lewis, H.E. and Masterton, J.P.: British North Greenland Expedition 1952-1954: medical and physiological aspects. *Lancet, ii:*494-500, 549-556, 1955.

Lloyd, R.M.: Medical problems encountered on British Antarctic expeditions. In Edholm, O.G. and Gunderson, E.K.E. (Eds.): *Polar Human Biology*. London, Heinemann, 1973, pp. 71-92.

Luft, U.C.: Principles of adaptation to altitude. In Yousef, M.K., Horvath, S.M. and Bullard, R.W. (Eds.): *Physiological Adaptations: Desert and Mountain*. New York, Academic Press, 1972, pp. 143-155.

Luft, U.C.: Aerospace environment. In Slonim, D.B. (Ed.): *Environmental Physiology*. St. Louis, Mosby, 1974, pp. 376-398.

McArdle, B., Dunham, W., Hollong, W.E., Ladell, W.S.S., Scott, J.W., Thomson, M.L. and Weiner, J.S.: The prediction of the physiological effects of warm and hot environments. *Med Res Council Rep No RNP47/391*. London, H.M. Stationery Office, 1947.

McCance, R.A., Ungley, C.C., Crosfill, J.W.L. and Widdowson, E.M.: The hazards to men in ships lost at sea 1940-1944. *Med Res Council Spec Rep No 291*. London, H.M. Stationery Office, 1956.

McFarland, R.A.: Psychophysiological implications of life at altitude and including the role of oxygen in the process of aging. In Yousef, M.K., Horvath, S.M. and Bullard, R.W. (Eds.): *Physiological Adaptations: Desert and Mountains*. New York, Academic Press, 1972, pp. 157-181.

MacInnis, J.B.: Living under the sea. *Sci Amer, 214(3)*:24-33.

Maegraith, B.G.: *Clinical Tropical Diseases,* 5th ed. Oxford, Blackwell, 1971.

Maher, J.T., Jones, L.G. and Hartley, L.H.: Effects of high altitude exposure on submaximal endurance capacity of men. *J Appl Physiol, 37*:895-898, 1974.

Manson-Bahr, P.H.: *Manson's Tropical Diseases,* 16th ed. London, Ballière, Tindall, and Cassell, 1966.

Marvin, U.B.: The moon after Apollo. *Technol Rev (MIT), 75*:3-13, 1973.

Matthews, B.: Limiting factors at high altitude in Peru. *Proc Roy Soc Ser B, 143*:1-4, 1954.

Mazess, R.B.: Exercise performance at high altitude in Peru. *Fed Proc, 28*:1301-1306, 1969.

Michel, E.L., Waligora, J.M., Horrigan, D.J. and Shumate, W.H.: Environmental factors. In Johnston, R.S., Dietlein, L.F. and Berry, C.A. (Eds.): *Biomedical Results of Apollo.* Washington, D.C., NASA, 1975, pp. 129-139.

Miles, S.: *Underwater Medicine.* London, Staples, 1966.

Miller, J.N.: Life support systems. In Bennett, P.B. and Elliott, D.H. (Eds.): *The Physiology and Medicine of Diving,* 2nd ed. London, Baillière Tindall, 1975, pp. 60-77.

Molnar, G.W.: Survival of hypothermia by men immersed in the ocean. *J Am Med Ass, 131*:1046-1050, 1946.

Monge, C.: Chronic mountain sickness. *Physiol Rev, 23*:166-184, 1943.

Moret, P.: Adaptation fonctionelle et métabolique du système cardio-vasculaire. Observations dan les Andes. *Bull Études Préhist Alpines, 9*:123-134, 1977.

Nier, A.D., Hanson, W.B., Shiff, A., McElroy, M.B., Spencer, N.W., Duckett, R.J., Knight, T.C.D. and Cook, W.S.: Composition and structure of the Martian atmosphere: preliminary results from Viking I. *Science, 193*:786-788, 1976.

Nuttall, J.B.: Emergency escape from aircraft and spacecraft. In Randell, H.W. (Ed.): *Aerospace Medicine,* 2nd ed. Baltimore, Williams and Wilkins, 1971, pp. 376-417.

Ohta, Y. and Matsunga, H.: Bone lesions in divers. *J Bone Jt Surg, 56B*:3-16, 1974.

Parker, D.E. and von Gierke, H.E.: Sound, vibration, and impact. In Slonim, N.B. (Ed.): *Environmental Physiology.* St. Louis, Mosby, 1974, pp. 119-162.

Peñaloza, D., Sime, F. and Ruiz, L.: Cor pulmonale in chronic mountain sickness: present concept of Monge's disease. In Porter, R. and Knight, J. (Eds.): *High Altitude Physiology: Cardiac and Respiratory Aspects.* Edinburgh, Churchill Livingstone, 1971, pp. 41-52.

Pollack, J.B.: Mars. *Sci Amer, 233(3)*:106-117, 1975.

Preston, F.S.: Medical aspects of supersonic travel. *Aviat Space Environ Med, 46*:1074-1078, 1975.

Prosser, C.L.: Perspectives of adaptation, theoretical aspects. In Dill, D.B.,

Adolph, E.F. and Wilber, C.G. (Eds.): *Handbook of Physiology*. Section 4. Washington, D.C., Am Physiol Soc, 1964, pp. 11-25.

Pugh, L.G.C.E.: Animals in high altitudes: man above 5000 metres — mountain exploration. In Dill, D.B., Adolph, E.F. and Wilber, C.G. (Eds.): *Handbook of Physiology*, Section 4. Washington, D.C., Am Physiol Soc, 1964, pp. 861-868.

Pugh, L.G.C.E.: High altitudes. In Edholm, O.G. and Bacharach, A.L. (Eds.): *The Physiology of Human Survival*. New York, Academic Press, 1965, pp. 121-152.

Pugh, L.G.C.E., Edholm, O.G., Fox, R.H., Wolff, H.S., Hervey, G.R., Hammond, W.H., Tanner, J.M. and Whitehouse, R.H.: A physiological study of channel swimming. *Clin Sci, 19:*257-273, 1960.

Rahn, H.: Lessons from breath holding. In Cunningham, D.J.C. and Lloyd, B.B. (Eds.): *The Regulation of Human Respiration*, Oxford, Blackwell, 1963, pp. 294-302.

Rambaut, P.C., Leach, C.S. and Leonard, J.I.: Observations in energy balance in men during space flight. *Am J Physiol, 233:*R208-212, 1977.

Rennie, D.W., Covino, B.G., Howell, B.J., Song, S.H., Kang, B.S. and Hong, S.K.: Physical insulation of Korean diving women. *J Appl Physiol, 17:*961-966, 1962.

Robinson, S.: Physiological adjustments to heat. In Newburgh, L.H. (Ed.): *Physiology of Heat Regulation and the Science of Clothing*. New York, Hafner, 1949, pp. 193-231.

Robinson, S.: Temperature regulation in exercise. *Pediatrics, 32:*691-702, 1963.

Robinson, S.: Cardiovascular and respiratory responses to heat. In Yousef, M.K., Horvath, S.M. and Bullard, R.W. (Eds.): *Physiological Adaptations: Desert and Mountain*. New York, Academic Press, 1972, pp. 77-97.

Rogers, A.F.: Antarctic climate, clothing and acclimatization. In Edholm, O.G. and Gunderson, E.K.E. (Eds.): *Human Polar Biology*. London, Heinemann, 1973, pp. 265-289.

Rogers, A.F. and Sutherland, R.J.: Clo values of polar clothing and their relation to "total number of layers" counts. *J Physiol, 240:*22p-23p, 1974.

Rummel, J.A., Michel, E.L. and Berry, C.A.: Physiological response to exercise after space flight. *Aerospace Med, 44:*235-238.

Saltin, B. and Hermansen, L.: Esophageal, rectal, and muscle temperature during exercise. *J Appl Physiol, 21:*1757-1762, 1966.

Samueloff, S.: Metabolic aspects of desert adaptation (man). In Levine, R. and Luft, R. (Eds.): *Advances in Metabolic Disorders*, Vol 7. New York, Academic Press, 1974, pp. 95-138.

Sauer, R.L. and Calley, D.J.: Potable water supply. In Johnston, R.S., Dietlein, L.F. and Berry, C.A. (Eds.): *Biomedical Results of Apollo*. Washington, D.C., NASA, 1975, pp. 495-515.

Sauer, R.L. and Jorgensen, G.K.; Waste management system. In Johnston,

R.S., Dietlein, L.F. and Berry, C.A. (Eds.): *Biomedical Results of Apollo.* Washington, D.C., NASA, 1975, pp. 469-484.

Sawin, C.F., Nicogossian, A.E., Schachter, A.P., Rummel, J.A. and Michel, E.L.: Pulmonary function evaluation during and following Skylab space flights. In Johnston, R.S. and Dietlein, L.F. (Eds.): *Biomedical Results from Skylab.* Washington, D.C., NASA, 1977, pp. 388-394.

Schaeffer, K.E.: Carbon dioxide effects under conditions of raised environmental pressure. In Bennett, P.B. and Elliott, D.H. (Eds.): *The Physiology and Medicine of Diving,* 2nd ed. London, Baillière Tindall, 1975, pp. 185-206.

Scholander, P.F.: Animals in aquatic environments: diving mammals and birds. In Dill, D.B., Adolph, E.F. and Wilber, C.G. (Eds.): *Handbook of Physiology,* Section 4. Washington, D.C., Am Physiol Soc, 1964, pp. 729-739.

Scholander, P.F., Hammel, H.T., Hart, J.S., LeMessurier, D.H. and Steen, J.: Cold adaptation in Australian aborigines. *J Appl Physiol, 13*:211-218, 1958.

Scholander, P.F., Hock, R., Walters, V. and Irving, L.: Adaptation to cold in arctic and tropical mammals and birds in relation to body temperature, insulation, and basal metabolic rate. *Biol Bull mar biol Lab, Woods Hole, 99*:259-271, 1950.

Severinghaus, J.W.: Transarterial leakage: a possible mechanism of high altitude pulmonary oedema. In Porter, R. and Knight, J. (Eds.): *High Altitude Physiology: Cardiac and Respiratory Aspects.* Edinburgh, Churchill Livingstone, 1971, pp. 61-68.

Shelton, W.R.: *Man's Conquest of Space,* 2nd ed. Washington, D.C., Nat Geograph Soc, 1972.

Shephard, R.J.: The athlete at high altitude. *Can Med Ass J, 109*:207-209, 1973.

Shephard, R.J.: Altitude training camps. *Br J Sports Med, 8*:38-45, 1974.

Shibolet, S.: Heat stroke: a review. *Aviat Space Environ Med, 47*:280-301, 1976.

Simmons, G.: *On the Moon with Apollo 15.* Washington, D.C., NASA, 1971.

Singh, I., Khanna, P.K., Srivastava, M.C., Lal, M., Roy, S.B. and Subramanyam, C.S.V.: Acute mountain sickness. *New Engl J Med, 280*:175-184, 1969.

Siple, P.A. and Passel, C.F.: Measurements of dry atmospheric cooling in subfreezing temperatures. *Proc Am Phil Soc, 89*:177-199, 1945.

Sirotinin, N.N. and Matsynin, V.V.: Adaptation to high altitude (Caucasus). In Collins, K.J. and Weiner, J.S. (Eds.): *Human Adaptability: a History and Compendium of Research.* London, Taylor and Francis, 1977, pp. 314-316.

Sloan, A.W.: *Physiology for Students and Teachers of Physical Education.* London, Arnold, 1969.

Sloan, A.W.: *The Physiological Basis of Physiotherapy.* London, Baillière Tindall, 1979.

Sloan, A.W. and Masali, M.: Anthopometry of Sherpa men. *Ann Hum Biol,* 5:433-459, 1978.

Smith, M.C., Heidelbough, N.D., Rambaut, P.C., Rapp, R.M., Wheeler, H.O., Huber, C.S. and Bourland, C.T.: Apollo food technology. In Johnston, R.S., Dietlein, L.F. and Berry, C.A. (Eds.): *Biomedical Results of Apollo.* Washington, D.C., NASA, 1975, pp. 437-468.

Spealman, C.R.: Physiological adjustments to cold. In Newburgh, L.H. (Ed.): *Physiology of Heat Regulation and the Science of Clothing.* New York, Hafner, 1949, pp. 232-239.

Spells, K.E.: The physics of the atmosphere. In Gillies, J.A. (Ed.): *A Textbook of Aviation Physiology.* Oxford, Pergamon, 1965, pp. 25-54.

Spencer, M.P. Decompression limits for compressed air determined by ultrasonically detected blood bubbles. *J Appl Physiol, 40:*229-235, 1976.

Steele, P.: Medicine on Mount Everest. *Lancet, ii:*32-39, 1971.

Strauss, R.H.: Decompression sickness. In Strauss, R.H. (Ed.): *Diving Medicine.* New York, Grune and Stratton, 1976, pp. 63-82.

Strughold, H.: The earth's environment and aviation. In Randel, H.W. (Ed.): *Aerospace Medicine,* 2nd ed. Baltimore, Williams and Wilkins, 1971a, pp. 22-34.

Strughold, H.: Circadian rhythms: aerospace medical aspects. In Randel, H.W. (Ed.): *Aerospace Medicine,* 2nd ed. Baltimore, Williams and Wilkins, 1971b, pp. 47-55.

Strydom, N.B., Benade, A.J.S. and van der Walt, W.H.: The performance of the improved microclimate suit. *J S Afr Inst Mining Metallurg, 76:*329-333, 1975.

Strydom, N.B. and Kok, R.: Acclimatization practices in the South African gold mining Industry. *J Occupat Med, 12:*66-69, 1970.

Strydom, N.B., Kotze, H.F., van der Walt, W.H. and Rogers, G.G.: Effects of ascorbic acid on rate of heat acclimatization. *J Appl Physiol, 41:*202-205, 1976.

Strydom, N.B., Wyndham, C.H., Williams, C.G., Morrison, J.F., Bredell, G.A.G., Benade, A.J.S. and von Rahden, M.: Acclimatization to humid heat and the role of physical conditioning. *J Appl Physiol, 21:*636-642, 1966.

Suzuki, M.: Thyroid activity and cold adaptability. In Itoh, S., Ogata, K. and Yoshimura, H. (Eds.): *Advances in Climatic Physiology.* Tokyo, Igaku Shoin, 1972, pp. 178-196.

Taylor, D.E.M.: Cold survival. *Brit J Sports Med, 6:*111-116, 1972.

Taylor, H.J.: High pressures, I. General. In Edholm, O.G. and Bacharach, A.L. (Eds.): *The Physiology of Human Survival.* New York, Academic Press, 1965, pp. 153-163.

Thompson, G.E.: Physiological effects of cold exposure. In Robertshaw, D. (Ed.): *Environmental Physiology II.* Baltimore, University Park, 1977, pp. 26-69.

Thornton, W.E. and Rummel, J.A.: Muscular deconditioning and its prevention in space flight. In Johnston, R.S. and Dietlein, L.F. (Eds.): *Biomedical Results from Skylab.* Washington, D.C., NASA, 1977, pp.

191-197.

Tikhomirov, I.I.: Cold and altitude stress. In Edholm, O.G. and Gunderson, E.K.E. (Eds.): *Polar Human Biology*. London, Heinemann, 1973, pp. 41-47.

Tobias, P.V.: Bushmen of the Kalahari. *Man*, 57:33-40, 1957.

Vanderveen, J.E.: Food, water and waste in space cabins. In Randell, H.W. (Ed.): *Aerospace Medicine*, 2nd ed. Baltimore, Williams and Wilkins, 1971, pp. 684-696.

Velazquez, T.: Aspects of physical activity in high altitude natives. *Am J Phys Anthropol*, 32:251-263, 1970.

Vorosmarti, J., Jr.: Saturation diving. In Strauss, R.H. (Ed.): *Diving Medicine*. New York, Grune and Stratton, 1976, pp. 287-301.

Waite, C.L., Mazzone, W.F., Greenwood, M.E. and Larsen, R.T.: Dysbaric cerebral air embolism. In Lambertsen, C.J. (Ed.): *Underwater Physiology*. Baltimore, Williams and Wilkins, 1967, pp. 205-215.

Walder, D.N.: The prevention of decompression sickness. In Bennett, P.B. and Elliott, D.H. (Eds.): *The Physiology and Medicine of Diving*, 2nd ed. London, Baillière Tindall, 1975, pp. 456-470.

Walder, D.N.: Aseptic necrosis of bone. In Strauss, R.H. (Ed.): *Diving Medicine*. New York, Grune and Stratton, 1976, pp. 97-108.

Waligora, J.M. and Horrigan, D.J.: Metabolism and heat dissipation during Apollo EVA periods. In Johnston, R.S., Dietlein, L.F. and Berry, C.A. (Eds.): *Biomedical Results of Apollo*. Washington, D.C., NASA, 1975, pp. 115-128.

Waligora, J.M. and Horrigan, D.J.: Metabolic cost of extravehicular activities. In Johnston, R.S. and Dietlein, L.F. (Eds.): *Biomedical Results from Skylab*. Washington, D.C., NASA, pp. 395-399.

Ward, J.S., Bredell, G.A.G. and Wenzel, H.G.: Responses of Bushmen and Europeans on exposure to winter night temperatures in the Kalahari. *J Appl Physiol*, 15:667-670, 1960.

Ward, M.: High altitude deterioration. *Proc Roy Soc Ser B*, 143:40-42, 1954.

Ward, M.: *Mountain Medicine*. London, Staples, 1975.

Weaver, K.F. Mariner unveils Venus and Mercury. *Nat Geograph Mag*, 147:858-869, 1975.

Webb, P.: Cold exposure. In Bennett, P.B. and Elliott, D.H. (Eds.): *The Physiology and Medicine of Diving*, 2nd ed. London, Baillière Tindall, 1975, pp. 285-306.

Weiner, J.S.: Climatic adaptation. In Harrison, G.A., Weiner, G.S., Tanner, J.M. and Barnicott, N.A. (Eds.): *Human Biology: an Introduction to Human Evolution, Variation and Growth*. Oxford, University Press, 1964, pp. 441-480.

Welch, B.E.: Ecological systems. In Brown, J.H.V. (Ed.): *Physiology of Man in Space*. New York, Academic Press, 1963, pp. 309-334.

Wells, C.L. and Paolone, A.M.: Metabolic responses to exercise in three thermal environments. *Aviat Space Environ Med*, 48:989-993, 1977.

Whedon, G.D., Lutwak, L., Rambaut, P.C., Whittle, M.W., Smith, M.C., Reid, J., Leach, C., Stadler, C.R. and Sanford, D.D.: Mineral and nitrogen metabolic studies. In Johnston, R.S. and Dietlein, L.F. (Eds.): *Biomedical Results from Skylab.* Washington, D.C., NASA, 1977, pp. 164-174.

Whiteside, T.C.D.: Motion sickness. In Gillies, J.A. (Ed.): *A Textbook of Aviation Physiology.* Oxford, Pergamon, 1965a, pp. 796-803.

Whiteside, T.C.D.: Visual mechanisms of special importance in aviation. In Gillies, J.A. (Ed.): *A Textbook of Aviation Physiology.* Oxford, Pergamon, 1965b, pp. 1004-1013.

Whiteside, T.C.D.: Visibility and perception. In Gillies, J.A. (Ed.): *A Textbook of Aviation Physiology.* Oxford, Pergamon, 1965c, pp. 1021-1028.

Whittingham, P.D.G.V.: Factors affecting the survival of man in hostile environments. In Gillies, J.A. (Ed.): *A Textbook of Aviation Physiology.* Oxford, Pergamon, 1965, pp. 479-513.

Williams, B. and Unsworth, I.: Skeletal changes in divers. *Australas Radiol, 20:*83-94, 1976.

Williams, C.G., Wyndham, C.H. and Morrison, J.F.: Rate of loss of acclimatization in summer and winter. *J Appl Physiol, 22:*21-26, 1967.

Wilson, O.: Experimental freezing of the finger: a review of studies. In Edholm, O.G. and Gunderson, E.K.E. (Eds.): *Polar Human Biology.* London, Heinemann, 1973, pp. 246-255.

Winter, D.L.: Weightlessness and gravitational physiology. *Fed Proc, 36:*1667-1671, 1977.

Wood, J.D.: Oxygen toxicity. In Bennett, P.B. and Elliott, D.H. (Eds.): *The Physiology and Medicine of Diving,* 2nd ed. London, Baillière Tindall, 1975, pp. 166-184.

Workman, R.D. and Bornmann, R.C.: Decompression theory: American practice. In Bennett, P.B. and Elliott, D.H. (Eds.): *The Physiology and Medicine of Diving,* 2nd ed. London, Baillière Tindall, 1975, pp. 307-330.

Wyndham, C.H.: Adaptation to heat and cold. In Lee, D.H.K. and Minard, D. (Eds.): *Physiology, Environment, and Man.* New York, Academic Press, 1970, pp. 177-204.

Wyndham, C.H.: The physiology of exercise under heat stress. *Annu Rev Physiol, 35:*193-220, 1973.

Wyndham, C.H. and Morrison, J.F.: Adjustment to cold of Bushmen in the Kalahari Desert. *J Appl Physiol, 13:*219-225, 1958.

Wyndham, C.H., Strydom, N.B., Benade, A.J.S. and van Rensburg, A.J.: Limiting rates of work for acclimatization at high wet bulb temperatures. *J Appl Physiol, 35:*454-458, 1973.

Yaglou, C.P.: Indices of comfort. In Newburgh, L.H. (Ed.): *Physiology of Heat Regulation and the Science of Clothing.* New York, Hafner, 1949, pp. 227-287.

Yaglou, C.C. and Minard, D.: Control of heat casualties at military training camps. *Arch Ind Health, 16:*302-316.

Young, A. and Young, L.: Venus. *Sci Amer, 233(3):*70-78, 1975.

INDEX

A

Aborigines, 43
Acceleration, 79, 80, 88, 93
  negative, 86
  positive, 82, 85
  transverse, 86, 87, 96
Acclimatization, 3
  (*see also* Cold, Heat, Mountains)
Acetazolamide, 75
Acidosis, 44
Adaptation (*see also* Altitude, Cold, Heat)
  behavioral, 3, 32, 43
  racial, 32, 42
Adipose tissue, 11
Adrenal
  corticosteroids, 75
  medulla, 10, 11, 41
Ainu, 42, 43
Air, 18, 19
  embolism, 60, 62, 84
Aircraft, 78-88
  cabin, 80, 82
  supersonic, 78, 83
Alacaluf Indians, 43
Aldosterone, 31
Altitude
  adaptation, 70
  supraterrestrial, 67, 90
  terrestrial, 67
Ama, 51, 54
Amharas, 74
Andes, 73
Anemometer, 12
Anorexia (*see* Appetite, loss of)
Antarctic, 37, 39
Antidiuretic hormone (ADH), 29, 31, 41, 75
Aortic body, 26, 70
Apollo, 90, 100

Appetite, loss of, 29, 68, 73, 75
Arctic, 37
Ascorbic acid (vitamin C), 36
Asphyxia, 66
ata units, 50
Atmosphere
  Earth, 18, 91
  Mars, 103
  Moon, 101
  space capsule, 94
  Venus, 103, 104
Auconquilca, 67, 73
Australia, 43
Aviation, 78-89

B

Balloon, 90
Barotrauma, 51, 53, 83
  ear, 60, 61
  lung, 60, 66, 84
  sinus, 61
Bathyscaphe, 59
Bends, 61, 84
Betamethasone, 75
Bicarbonate, 24, 45, 70
Black-out, 85
Blood
  flow (*see* Circulation)
  oxygen capacity, 22, 71
  oxygen saturation, 21, 72
  pressure, 52, 70, 71
  viscosity, 29, 34, 46, 69, 71
  volume, 31, 71
Bohr effect, 23
Bone
  aseptic necrosis, 65
  decalcification, 94, 98
Boyle's law, 50, 79
Buddy system, 57
Bushmen, 32, 43, 44

127